ADHESIVES AND SEALANTS

Adhesives and Sealants

David Lammas

ARGUS BOOKS

Argus Books
Argus House
Boundary Way
Hemel Hempstead
Hertfordshire HP2 7ST
England

First published by Argus Books 1991

© David Lammas, 1991

ISBN 1 85486 048 8

Phototypesetting by Photoprint, Torquay, Devon
Printed and bound in Great Britain by Biddles Ltd, Guildford and King's Lynn

CONTENTS

ADHESIVES AND SEALANTS

PREFACE

There still seems to be a reluctance, especially among some of the older generation, to use the word 'adhesive'; 'glue' is the word they were brought up on and still prefer to use. However, if we look up the word 'glue' in a dictionary we find it means 'hard gelatin got from hides and bones, used warm as a cement'. This is rarely what people mean nowadays when they speak of glue, and the term has come to be used in a general sense for almost any substance that will stick things together.

When we look up the word 'adhesive' it is defined as 'sticking' or 'sticky', and 'adhere' means 'stick firmly'. An adhesive, then, is merely a substance that will stick things firmly together; in other words, exactly what many people mean when they say 'a glue'.

A 'sealant' is also usually sticky, though it often has less strength than adhesives intended for fastening things together. Its function is only to prevent leakage of liquids or gases through joints that are secured by rivets, bolts, threads and so on. A sealant may remain permanently sticky or tacky, or it might set yet remain flexible enough to prevent cracking with consequent leakage. Some adhesives act as both fasteners and sealants, so it is not that easy to draw a rigid dividing line between them.

An engineering adhesive is one that is employed to fasten together mechanical components or assemblies on a permanent or semi-permanent basis – the latter referring to parts that may need to be dismantled or adjusted at some future time. A structural adhesive is, as the name implies, one that is used to build engineering structures, for example aircraft fuselages and wings. The materials to be joined are usually sheet metals, plastics or composites. Indeed it is in the field of aircraft construction that structural adhesives first found extensive use, probably as a result of the experience gained with laminated wooden parts. Weight reduction, plus the time saved by eliminating numerous rivetting operations coupled with a smoother skin and less likelihood of corrosion, were important factors in its adoption.

Other industries were at first slow to appreciate the advantages of adhesive assembly. It was looked upon merely as a means of utilising out of tolerance

7

components that would otherwise have been scrapped at one time, but now it is true to say that there are few areas where it is not found to some extent. Today designers specify from the outset the joint configurations and tolerances most suitable for the technique. This frequently allows automated adhesive application and assembling without interrupting the flow of parts. Since threads, keys, pins, rivets, screws and the like are thereby rendered largely unnecessary, in many instances a multitude of machining operations can be dispensed with leading to increased efficiency with lower costs.

It must not be thought that such joints are simply a cheap substitute for what was formerly considered 'a proper job'. Extensive testing over the years during which the method has been used show that a well-designed adhesive joint can be superior to a conventional joint in terms of tensional, torsional and fatigue strength, as well as preventing fretting or corrosion between the mating parts.

There is now enough evidence of their successful application to model engineering problems to convince even the most sceptical of us that the modern techniques of adhesive assembly and sealing have an important place in the amateur workshop. I hope that this handbook will prove useful in designing suitable joints for items of workshop equipment, as well as for models of all types where the methods are applicable. It is also intended as a guide in selecting the most suitable products to use for different materials and under differing conditions.

CHAPTER 1

TRADITIONAL GLUES

Until quite recently, the glues available were derived solely from natural sources, being used as found or with relatively little pre-treatment of the ingredients to convert them into the desired form.

Substances that were naturally sticky included the resinous exudations of certain trees, notably the pines, egg white, some waxes, and the so-called mineral bitumen found oozing from the ground in various parts of the world. This latter substance represents remnants of former oil deposits that have escaped to the surface as a result of earth movements and erosion, the lighter fractions having long ago escaped by evaporation. Bitumen is, therefore, in reality a product of the decomposition of organic matter.

Another tree product was the sap of the rubber tree *Hevea Brasiliensis* which originally was found only in the Brazilian jungles. The milky sap flowing from wounds in the bark of the tree is known as rubber latex and is an example of an emulsion. It consists of very fine droplets of liquid rubber suspended in water. When the water evaporates the droplets fuse together and dry into a translucent film of pure

rubber, sticking firmly to any material in contact with it. Another example of an emulsion is milk composed of fine droplets of butterfat suspended in water.

Once the process of grinding cereal grains to make flour had been discovered, it cannot have taken long to observe that damp flour formed a rather sticky paste that would become solid when dried. Older readers may remember their parents or grandparents preparing flour and water paste as a simple household adhesive. One widespread use for it was the hanging

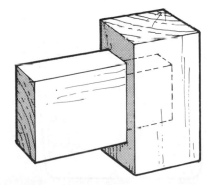

Typical woodwork joint.

9

of wallpaper. As it was made from a food product this fact did not escape the attention of a variety of bugs, bacteria and fungi particularly when, as in many old houses, the walls became damp.

For centuries, the furniture trade relied on a glue of animal origin. When bones or hides are boiled in water, the liquid thickens. If the water is then evaporated away, a residue of gelatin is left familiar in its purified form as table jelly. The cruder product used as glue was a hard, brown, toffee-like substance that had to be broken into small pieces with a hammer before making up the glue, although it was sometimes available as grains or powder.

As the ready-to-use glue did not keep, it was necessary to prepare it fresh for each job. The prescribed ritual involved soaking the broken pieces in cold water overnight causing them to soften and swell up. The excess water was poured off and the wet lumps placed in a glue-pot. This was, in reality, two vessels – an inner one to contain the glue and an outer container partly filled with water. The pot was heated to boil the water which, in turn, heated the lumps of glue until they became liquid. The double vessel prevented overheating which

Traditional glue-pot outer vessel. Inner one has unfortunately not survived.

would decompose the gelatin resulting in loss of strength. The liquid glue was applied to the joint surfaces hot and the joint closed before it had cooled. The whole thing was then clamped firmly together for a period sufficient to allow complete absorption of the water into the wood and its consequent evaporation.

The strength of a properly made gelatin-glued joint is greater than that of the wood itself; in other words, the wood will break before the joint comes apart. A disadvantage of the glue is that it readily softens and weakens its grip if allowed to become damp. Again, being a food product it is attacked by living organisms to the detriment of its strength, but this is only likely when

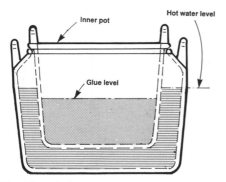

Traditional glue-pot.

in a damp condition. The property of softening in contact with water can be put to good use where the renovation or repair of valuable old furniture is concerned, for instance to re-glue loose joints or to remove a broken piece for repair or replacement. A very effective method is to play a jet of steam over the area of the joint – the combination of heat and dampness soon softens the glue sufficiently to ease the joint apart. A short soak in warm water removes all adhering glue, leaving a clean surface for re-fixing. Mortice and tenon joints where the glue-line is largely inaccessible are tackled by first drilling a hole about ⁵⁄₁₆″ diameter down to the joint surface from a position where it is unlikely to be noticed on the finished job. A ¼″ pipe connected to the kettle spout or other source of steam directs it right to the heart of the joint. The hole is of course carefully plugged and finished to match the original surface afterwards. For this reason, as well as to ensure authenticity, it is still considered good practice to use animal glue in the antiques business.

A traditional glue with which everyone is familiar must be gum arabic, and this represents one of the earliest 'instant glues'. Found on envelope seals and postage stamps it only needs a quick lick and immediate pressing together of the parts to take a very firm hold. Who has not sealed the envelope without first putting the letter inside? In such a case there is very little time in which to reopen it without going to the trouble of steaming.

Gum arabic illustrates another useful feature in that the two parts of the joint do not adhere unless wetted. The mat-

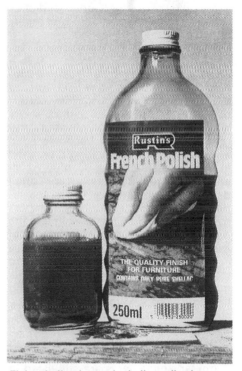

Flake shellac (centre), shellac adhesive solution in alcohol (left) and commercial solution (right).
Note: Rustin's Shellac Varnish contains a higher percentage of shellac than the French Polish shown.

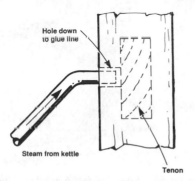

Hole down to glue line

Steam from kettle

Tenon

Steaming a joint to soften the glue.

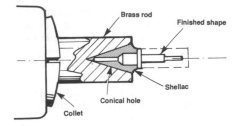

Watchmaker's wax-chuck.

erial is not sticky in its dry state, unlike some of the newer press-and-stick adhesives which must be kept apart until the moment of sealing.

Shellac is another very useful adhesive of ancient development. It is a dried secretion found upon certain subtropical bushes that have been host to the Lac insect. The brown, shiny, water-resistant solid melts to a thick sticky liquid when moderately heated, and upon cooling it again becomes solid and firmly stuck to anything it touched in the liquid state. Long used by watch and instrument makers where small delicate components were to be machined, it had advantages over complex mechanical holding methods in that the part was not marked or distorted out of truth.

Two particular methods of holding work in the watchmaker's lathe in this manner may be found by looking through old watchmaking books, although the holders were often called wax-chucks. The first type resembled a very small faceplate being made in several sizes from about 7mm to 40mm diameter; these items are still listed in watch- and instrument-maker's equipment catalogues. Flat components like gearwheels or short levers, etc, would be held on these for facing, recessing and boring. The procedure was to warm the chuck in the flame of a small spirit lamp until a piece of shellac, touched to the surface, melted forming a thin film of the adhesive. The part was quickly pressed onto the chuck and roughly centred by eye. With the work spinning in the lathe, final accurate centring was accomplished by the simple expedient of holding the pointed end of a toothpick against it until the shellac had cooled and set.

For mounting the larger work in this manner a number of favourite recipes were recommended consisting mostly of powdered material such as 'brick

Wax-chuck on watchmaker's lathe and glass alcohol lamp used to warm it.

Small gear wheel mounted on wax-chuck with melted shellac for boring operation.

dust' which was stirred into melted shellac; on cooling such mixtures would set rock hard. A little shellac goes a long way since the adhesive can be re-melted over and over again. The component is removed by applying gentle heat then remaining traces dissolved away in warm methylated spirit or industrial alcohol.

The second method, for holding tiny watch balance staffs or spindles, used a different form of wax-chuck. This was made by holding a short length of brass rod in the lathe collet then boring by hand a true conical depression in the end. The partly turned staff was, by virtue of its shape, impossible to hold in a collet or mechanical chuck, but could be easily held in the conical hole by a dab of melted shellac. The inner end was perfectly centred against the cone

while the outer end could be adjusted to run true as before. Since small parts like this were usually turned from hardened steel tempered to a blue colour (spring temper), the strength of the adhesive can be appreciated. I would recommend that the model engineer obtains a small quantity of flake shellac; its use can solve many little holding problems.

This versatile substance can also be used in another way. When dissolved in alcohol a sticky varnish results which was formerly much in demand for wood finishing, used either as a straight varnish or as the basis for French polishing. When sufficiently thick this solution can be used as an adhesive, since the alcohol evaporates rapidly from porous surfaces leaving a hard, tightly adherent film of shellac within

13

Another type of wax-chuck held in lathe collet. Tiny spindle fixed in chuck by melted shellac for turning operations.

the joint. In the early days of the electrical industry shellac varnish was widely used as an insulating and water-resisting coating, while the adhesive properties were put to good use on threaded wire connections and adjusting screws to prevent inadvertent loosening, to give visible evidence that inspection and testing had been carried out, and to discourage unauthorised tampering. A coloured dye was often incorporated in the mixture enabling the sealant to be readily observed. Modern threadlocking adhesives are not, then, a new development, merely an extension of an old idea.

Waxes are another form of natural adhesives that used to be available in a number of varieties. You only have to think of beeswax, paraffin wax,

carnauba wax, for instance, to realise that the differing hardnesses, melting points and solubilities in several types of solvent constitute a useful range of properties.

The gemstone cutter for example mounted the rough pebble on the end of a short 'dop stick' with melted wax. This method of fixing gave a waterproof hold, did not damage the stone or cause obstruction during the grinding, and also enabled the stone to be readily removed for repositioning. The actual grinding was done by holding the mounted stone at a fixed angle against a revolving disc fed with a thin slurry of powdered abrasive and water. The water kept the work cool, hence the need for a waterproof wax.

We have already mentioned the word

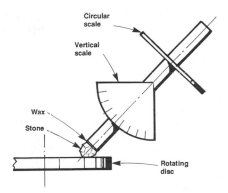

Circular scale

Vertical scale

Wax

Stone

Rotating disc

Gemstone facetting.

gum which originally meant a vegetable-derived substance that became sticky when wet, for instance gum arabic. Another example of a gum not designed as such, but sometimes knowingly or unknowingly employed

as an adhesive, is chicle, the sticky component of chewing gum. It has, I understand, been used for all manner of emergency repairs including the sealing of leaky petrol tanks, and its adhesion to bedposts and carpets is legendary. Nowadays the word gum has a somewhat looser definition, gum rubber does not become sticky when wetted with water but it does do so when certain solvents are involved. It was at one time the practice to dissolve old worn out crêpe rubber shoe soles in petrol to make a very serviceable adhesive. This was possible since pure natural gum rubber was the material from which the soles were made; a vulcanised rubber would not dissolve in this way. When petrol was five bob a gallon and shoes cost a few shillings this was a cheap adhesive.

An example of an early glue manu-

The traditional method of thread-sealing with hemp strands and Boss White.

15

factured using a chemical process to modify a natural substance is casein or milk protein. Surplus milk was soured by adding chemicals, then the curds were treated to remove the butterfat and soluble minerals before a further chemical treatment and drying of the casein to a powder form. This product would be mixed with a small amount of water to a pasty consistency immediately before use. Casein-based adhesives are still manufactured for wood bonding, mainly cheap plywood, and as a tile fixative for interior use. It is more damp resistant than animal glue but is not intended for permanently damp situations or exterior application.

So far we have noted a number of adhesives that were available in different forms and worked in different ways, each was also intended to be used with only a limited number of materials. This leads us to one very important point that needs to be clearly understood before you can attempt to select an adhesive for a particular job.

There is NO one universal adhesive as yet. Some may be described as multi-purpose or have the word universal in their name, but this does not mean they will stick every type of material.

Taking a look now at some of the older sealing compounds, we can start with one that is still with us. Those of you who have dabbled or perhaps worked in the field of plumbing when screwed iron fittings were the order of the day, with old steam engines, water cooled stationary IC engines, or central heating systems, will know what a fine mess you can get into handling plumbers' jointing, which for some obscure reason goes under the name of boss white.

The gooey white paste was applied to the male threads then several strands of hemp fibres were wound on to help fill the clearance and reinforce the paste, enabling it to withstand the pressure without being forced out of the joint. If you got it wrong, the whole lot just unwound as the parts were being screwed together necessitating complete cleaning off and a fresh start, hence the mess. Composed of linseed-oil mixed with a white powder, the sealant gradually hardened in the joint as the linseed-oil became chemically changed through reaction with oxygen from the air. This change is identical to that which occurs when oil-based paints harden; the initial tacky paint film left after the solvents have evaporated quickly hardens on reaction with oxygen. The action is so quick in this case because a large area of paint is directly exposed to the air and the film is very thin. Within the confines of a screwed joint the oxygen takes much longer to penetrate.

It will be appreciated that besides preventing leakage, the sealant also acted to reduce corrosion of the threads; even joints made years ago can sometimes be unscrewed to reveal near perfect thread condition.

Where flanged joints were concerned it was the usual practice to fit some sort of compressible gasket material between the faces, although in many types the compressibility was only slight. Such 'give' in the gasket was essential in taking up any slight imperfections in the joint surfaces as well as allowing for minute deviations from parallelism as the bolts were tightened.

With low pressures quite soft materials like felt or canvas could be used in conjunction with a good smear of red or white lead oil bound paste. As pressures increased firmer materials including asbestos or fibre reinforced

Groove filled with damp mixture of sulphur, ammonium chloride and iron filings

Rust seal.

their own, or with perhaps a smear of grease to help hold them in place while the parts are fitted together, but only so long as the joint surfaces are flat and parallel without dents or scores across the face. When distortion or damage has occurred it becomes essential either to do the job properly and reface the joint or to resort to those sticky sealants generally referred to as gasket goos. Such substances have been around for a very long time in one form or another.

A rather crude, but nevertheless effective, method of sealing semi-permanent joint faces on some old steam locomotive cylinders was the 'rust seal'. A shallow groove was made all around the joint face inside of the line of bolts and immediately before closing the joint this groove was filled with a paste of sulphur, fine iron filings and 'sal ammoniac' or ammonium chloride. Being rather like ordinary salt you can imagine it was not many hours

compounds were developed; present day steam sealing gaskets are of this type.

The paper gaskets that prevent oil leaks from small gearboxes and other mechanisms are perfectly adequate on

Hand bearing compass repaired by sealing with Litharge-Glycerine poylmer 30 years ago.

17

before the filings had turned to a mass of rust. The sides of the groove also rusted somewhat so that it took a good hold. Because the volume of rust is much greater than the volume of iron that produced it, a considerable expansion took place within the groove forming a very tight seal indeed.

Another sealant for smaller metal components consisted of phosphoric acid mixed with copper oxide. This would form thin even joint lines between flat surfaces of steel that were extremely difficult to separate.

One old recipe that I once used to seal the glass window of a hand bearing compass to the brass body was a mixture of litharge (yellow lead oxide) and glycerine; this was quite successful for preventing evaporation of the alcohol filling the compass bowl. It is an interesting example of a sealant made from a mineral component and an organic substance where chemical reaction between them caused it to set and adhere. It is really a two-component chemical set adhesive similar in action to the epoxy resin types in that a form of polymerisation takes place.

A sealant that was at one time used in chemical apparatus and early electrical equipment is ordinary solid sulphur. When melted it adheres well to metals and glass forming a tight seal and is, in addition, a very good electrical insulator.

A more modern electrical sealing and waterproofing mixture was Henleys compound made of castor oil mixed to a putty with a powdered filler such as china clay. It was similar to plasticine which consists of china clay mixed with a grease like vaseline and both of these substances remain perpetually soft. Although they do soften more as the temperature rises, and stiffen as they become cooler, they never 'set' solid.

Chattertons compound was yet another sealant with waterproofing properties and was sold in sticks about one inch diameter by 6 inches or so in length. This was a hot-melt stick made from various resins and fillers.

Stockholm, or wood, tar as well as coal tar and bitumen, pitch or asphalt could be used alone or in admixture with other ingredients and fillers to give a range of sealants exhibiting differing physical properties.

We can see therefore that our forebears were not lacking in inventive genius or slow to adapt natural substances to the needs of the day. Periods of great upheaval and change such as the industrial revolution merely stimulated the development of new products and applications.

CHAPTER 2

THE NATURE OF THE ADHESIVE JOINT

Before you can begin to design a joint and select a suitable adhesive for it, some understanding of material surfaces and the manner in which adhesives bond to them is required, together with an appreciation of how they react to mechanical stresses.

It is perhaps best to start by considering the more familiar methods of joining materials to find out how the adhesive process differs from them, how it compares with them for strength, what are the advantages and disadvantages, and so on.

With wooden structures the components may be nailed, screwed or bolted together. The first relies on the frictional grip of the wood around the nail, the second is really a spiral wedge action, while the third represents a compression joint. They all act over a very limited area of the wood even when the fixings are placed reasonably close together. Whether the nails or screws are made of brass or steel it is obvious that they are in themselves much stronger than any adhesive, so how is it then that wooden structures can be successfully glued together? The answer lies in the fact that adhesive bonding makes use of the whole of the joint area rather

than just isolated points. It should be remembered that we do not expect nails or screws to hold well in the end-grain of timber, neither should we expect a strong adhesive bond to result from such practice – the joint must be designed to ensure adhesion parallel to the grain.

For metal structures, particularly of steel, there is a wide choice of fastening methods involving spot-welds, fusion seam welds, brazing, rivetting, bolting, etc. Spot-welding requires expensive equipment, the small area joints lead to high stresses, the open seam is susceptible to corrosion, and joints must be designed for ease of access rather than for optimum strength. Fusion welds again need special equipment, since the high temperature affects the properties of the adjacent metal, often resulting in distortion and shrinkage stresses. Brazing, too, suffers from these drawbacks. Bolts and rivets require accurately drilled holes and additional components. Corrosion can also be a problem.

By contrast, adhesive-bonded joints may be designed for best performance. Setting or hardening of the adhesive takes time but usually takes place at room temperature thereby eliminating distortion and residual stresses. Since the whole joint area is bonded, service stresses are more widely distributed and a much stiffer structure is obtained. This complete coverage seals the joint against leakage or the entry of moisture thus overcoming the corrosion problem.

The disadvantages of adhesives are their relatively low strength, softening and weakening at temperatures as low as 100 to 200 degrees Celsius, and long term deterioration, particularly as the temperature rises. Many adhesives are weakened by exposure to water and some solvents. They tend to fracture under sudden shock loads as well as exhibiting poor resistance to loads tending to peel the adhesive from the substrate.

Considering now the nature of the joining methods, it is well known today that the strength of a metal, for instance, depends upon the forces of attraction between individual atoms making up the bulk of the metal. When the metal fractures, these chemical bonds joining atom to atom are being broken by the million. Such bonds can only be remade by melting the metal in order that the atoms may move about freely to become again closely aligned in an orderly crystal structure as it solidifies on cooling; this is what occurs during fusion welding. During hammer welding of iron the metal is brought almost to fusion temperature, in which state it becomes quite plastic. The heavy blows then force the surfaces into close contact so that the material recrystallises across the joint line to become practically homogeneous.

An example of peel separation. Aluminium strips with epoxide adhesive.

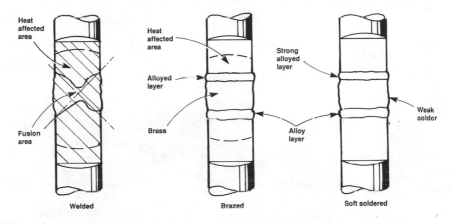

Mild steel rods joined in various ways.

In a brazed joint the temperature is high enough for the molten brass to dissolve iron atoms from the surface to give a diffuse joint line rather than a sharp boundary. This partial alloying at the surface can only occur between certain metals where the one is 'wet' by the other. A result of the alloying is that the actual region of mixing becomes harder and stronger than the brazing metal itself. A similar action takes place during soft soldering, the wetting of the surface then being referred to as 'tinning'. Again, this very thin layer is modified by alloying with the surface to become stronger than the parent solder, which is why any excess of weak solder should be squeezed from the joint. A close look at the copper 'bit' of a well-used soldering iron will show the extent to which copper is dissolved by molten solder at a temperature far below the melting point of copper.

When an adhesive-bonded joint is compared with these other forms of fastening it should be obvious that it will be much weaker than welding since the polymers from which adhesives are made are much less strong than metals. Similarly we should not expect an adhesive bond to be as strong as an equal area of brazed metal. Therefore to get some sort of perspective on the problem it can be said that an adhesive joint is more nearly comparable to a soft soldered one, although even the best adhesives are inferior in those cases where the load tends to peel the surfaces apart from one end.

To understand the manner in which adhesives become attached to the surface of the substrate it is informative to look at another phenomenon — there can be no question of the substrate dissolving in the glue in the case of metals so just what does happen?

A set of engineer's slip gauges consists of small flat pieces of very hard steel that have been accurately ground and polished to precise thicknesses for use in precision measuring operations. The surfaces are flat to almost one millionth of an inch (about one ten-thousandth of a millimetre) which we may fondly imagine to be perfectly flat and smooth. When two slips are placed

21

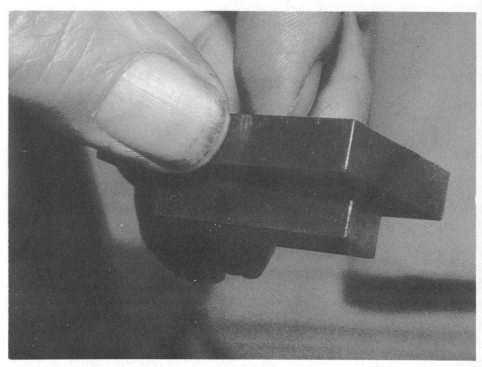

Photograph illustrating the adhesion of finely ground and polished metal surfaces. (Photo courtesy Ted Jolliffe)

face to face and twisted into close contact it is found that they adhere together so well that appreciable force is required to separate them again. It was at one time thought that the adhesion was due to the forces of attraction between the atoms of metal of each surface, it is after all such forces that hold the solid metal together. One reason why that explanation does not find favour today is because interatomic forces are extremely short range and are only effective over distances comparable to the radius of an atom. Like magnetic attraction the strength falls off rapidly as the distance apart increases.

Unless atoms are within 'touching distance' then the forces are quite feeble and when the surface roughness of slip gauges is looked at on an atomic scale they are far from smooth or flat. The radius of an atom as a round number can be put at one ten-millionth of a millimetre – we saw that slips are flat to about one ten-thousandth of a milli-metre, which means that the surface is composed of hills and valleys about one thousand times greater than the radius of an atom.

Remembering that two rigid surfaces can only touch each other at three sepa-rate points, in simple terms the rest of

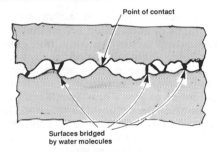

Point of contact

Surfaces bridged
by water molecules

Adhesion of slip gauges.

the area is a gap that varies from a minimum where two peaks are adjacent to a maximum where two valleys are opposite each other. The number of points at which atoms on opposing faces are close enough together to experience a significant attraction is seen to be rather limited. This assumes also that the surfaces are perfectly clean and free from other substances which is extremely unlikely on that sort of scale.

It has been found that if the slips are put into a very high vacuum, such as exists in space, they do not adhere together giving additional evidence that it is not a case of direct attraction. The effect of the vacuum is to 'boil off' gaseous and liquid molecules from the surfaces so it would appear that they must have played some part in sticking the pieces of metal together. It is a fact that many surfaces have a very thin layer of water molecules attached to them even in summer when they appear to be perfectly dry since water, or water vapour, is always present in the atmosphere. On suitable surfaces this water layer is not just sitting there to be readily shaken off but forms a type of chemical bond with the surface atoms, which is why extreme measures like the high vacuum are needed to remove it.

When slip gauges stick together, it is the microscopic layer of water molecules that bridges the gaps with chemical bonds. They cannot bridge the widest parts so it is only a partial coverage of the area giving a limited grip. If a substance having similar chemical properties to water could be placed between the surfaces in a liquid state so as to thoroughly wet them and fill all the gaps, and then set to a hard strong solid, the bond would be much stronger. This is indeed what happens when two surfaces are glued together.

Incidentally, in the field of lubrication the next generation of lubricants will consist not of oil, which has very poor lubricating properties, but of liquid chemicals having the ability to form strong bonds to the metal surfaces; since it remains liquid the parts will not be stuck together but the surface film prevents the metal-to-metal contact that causes wear.

It will be appreciated that not all types of chemical bond have the same degree of strength, it depends upon the actual sorts of atoms being joined. This is why an adhesive may be successful in sticking one material but fail to adhere to another; it is rather like shaking up oil and water. In the case of the latter, they just do not mix or it can be said that there is a lack of affinity. This can be seen when an attempt is made to glue polythene which is regarded as being one of the more difficult plastics. The analogy with oil and water is particularly apt here because many adhesives contain hydroxyl groups, as in water, which can be represented as HOH, or hydrogen attached to the hydroxyl OH, and polythene is a hydrocarbon similar in structure to oil.

The number and variety of adhesives that have been developed is a result of

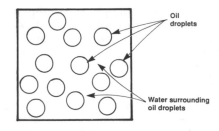

Oil in water emulsion.

these differences in the properties of the substrate materials. There is as yet no single universal glue that will stick all substances, and indeed it is rather unlikely that such a product can be found.

Those materials like polythene which are difficult to bond are made from types of chemical that do not readily undergo chemical reactions – they are said to be inert or inactive. We do know however that polythene will burn, which is a reaction between the carbon and hydrogen of the polymer with the oxygen of the atmosphere. It is an oxidation reaction that proceeds to completion because of the high temperature involved, the final products being carbon dioxide gas and water vapour.

Various methods of activating the surfaces of such polymers so that they will accept adhesives have been used for many years but are not always easy to apply. These surface activation processes rely on a partial or controlled oxidation of a very thin layer leaving the interior of the polymer untouched. The three common ways of doing this are:

(1) By treating the surface with powerful chemical solutions. In the original application natural rubber was treated with concentrated sulphuric acid for a few minutes then thoroughly washed and dried. A modified finely crazed surface was formed. More recent treatment of other polymer types usually involves chromic acid solutions. It will be appreciated that this can be a rather hazardous process more suited to industrial conditions than for the amateur.

(2) By briefly passing a flame across the surface. It requires a small clean gas flame to be moved across the material at a rate sufficient to produce a slight shine on the surface without melting the polymer. Oxidation and 'free radical' formation provide attachment sites for the adhesive molecules which must be applied immediately following treatment.

Surface activation of polythene strip by briefly passing a gas flame across the surface.

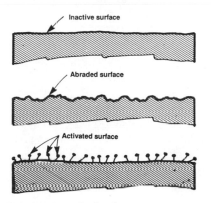

Inactive surface

Abraded surface

Activated surface

Surface treatment of polymers.

(3) By passing an electrical 'brush dis charge' over the surface. A brush

discharge consists of a flow of ionised gases produced around a wire carrying very high voltage, in the order of ten thousand volts, in air. It is these energetic ions that react with the polymer surface to form reactive sites in a similar manner to the flame technique.

The equipment required is rather complex, it must be well guarded and screened being more suited to industrial usage. It has the potential of easier control, but for all that the flame process is widely used, especially as a preparation for printing or labelling plastic bottles, the automated set up handling thousands of items an hour.

CHAPTER 3

JOINT DESIGN

Many woodwork joints have been evolved over the years that are capable of withstanding considerable stresses even when no glue is present. In the days when waterproof glues were unknown this was a necessary requirement for outdoor woodwork of all kinds, ranging from house building to wheelmaking. Sometimes fastenings in the form of pins or wedges prevented movement but in the main, thought was

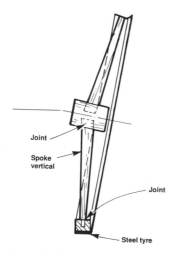

A wooden wheel.

given to the likely direction of the forces to be resisted and an appropriate type of joint accordingly selected.

The wheel is a good example to start with. Consider how the spokes were morticed into the nave, in a simple wheel for light duty there might be a single row of spokes all aligned at right angles to the axle. This wheel would be satisfactory only if the load acted directly downwards through the spokes, but as soon as it turned a corner or moved over uneven ground the joints would experience a sideways force leading to rapid loosening. Wheels designed for heavy work often have two rows of spokes set slightly apart at the nave but converging towards the rim in order to accommodate such stresses.

The joint then is designed to hold the spoke at the correct angle so that the load can be taken in compression, the load holds the joint together while the tyre prevents the whole thing from falling apart.

What, if anything, would be the advantage of using an adhesive in such a joint? While it would not add to the load carrying capacity of the wheel, an adhesive would stiffen the joint against

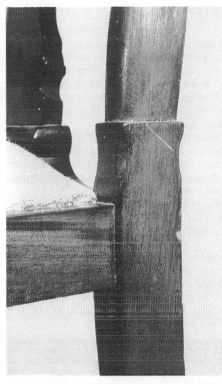

Seat to back mortice and tenon joint of chair.

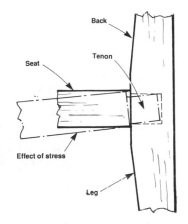

Glue failure in chair.

forces tending to loosen the fit and would also seal the joint preventing rot. It is interesting to note that when joints became loose in dry weather it was necessary to stand the wheels in water to swell the wood and tighten them up again, a procedure that can only have encouraged eventual decay.

The joints of a chair too are largely in compression so long as we sit still and upright. Leaning back in a chair places severe racking strains on the joints between seat and rear legs — in the absence of glue these mortice and

tenons soon enlarge the gap creating a wobbly perch. Even with glue, unless the joint is well made in the first place then thoroughly coated with hot glue and quickly clamped together until set, such treatment can lead to early failure.

Pull out the drawer of an old, pre-chipboard, piece of furniture and you will see a dovetail joint between drawer sides and front taking the tension stress. Again, this type of joint will resist the expected forces without gluing but glue is used to hold the parts together and stiffen up the joint against other unintended stresses.

We are now in a position to examine the forces acting on simple types of adhesive joints and to appreciate how they affect the glue line. Instead of wood, with its directional properties, more simple metal components will suffice.

The three basic ways in which a joint can be loaded are in *tension*, in *compression* and in *shear* or, to put it in more conventional terms, *pulling,*

27

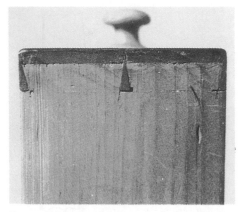

Dovetail joint in old drawer where animal glue was used.

squashing and *sliding* respectively. Of these, adhesives tend to be good in compression and shear but rather less so in direct tension. Dealing with each in turn we have:

(1) Direct Tension

This is the case of a simple butt joint where the load is tending to separate the two parts at right angles to the plane of the adhesive. The strength of the joint depends on (a) how well the adhesive sticks to the substrate and (b) the tensile strength of the cured or set adhesive.

Bearing in mind that modern adhesives are polymeric substances we would expect the strength to be similar to that of some plastics, i.e. less strong than metals.

Where it is desired to construct a joint between two metal parts in tension, allowance must be made for the difference in strength of the particular metal section and the adhesive selected. If the metal is ten times stronger than the adhesive, it is apparent that the area of the joint surface must be ten times greater than the cross sectional area of

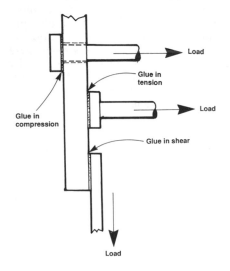

Types of stress.

A direct tension joint made between the heads of two bolts using Loctite 601 retainer. Although not a true adhesive it gives a strength of about 1000 psi.

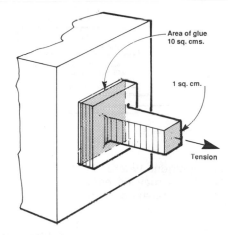

Area of glue
10 sq. cms.

1 sq. cm.

Tension

Increased glue area.

the metal to obtain equivalent strength. For this reason tension joints tend to be rather bulky in some applications. They do have more serious drawbacks in that stress concentrations at the edge of the bond can lead to fracture if the pull is not exactly at right angles, particularly under shock loading; a crack starting at the edge quickly spreads through the adhesive layer in a manner known as cleavage failure. Design should ensure that a sufficient margin of safety is allowed to minimize this danger. It may be said here that if examination of the failed bond surface shows the adhesive to have pulled cleanly away from the surface, it is probable that either it was not properly prepared in the first place

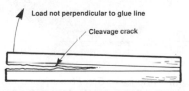

Load not perpendicular to glue line

Cleavage crack

Cleavage failure.

or the wrong adhesive was selected. In a good joint the adhesion to the surface should be greater than the tensile strength of the adhesive itself so any fracture should occur within the glue.

(2) Compression

This again can be a butt joint where the load is tending to press the two surfaces together at right angles to the plane of the adhesive. The function of the glue in this case is mainly to locate the parts, while the strength relates to the ability to resist compression which, for adhesives, is generally good. Again, design should endeavour to provide adequate joint area to accommodate stresses away from the perpendicular liable to cause cleavage. It is often possible to change an initial design by quite simple modifications so that suspect tension joints become more predictable compression joints. This might be one way of altering the joint configuration in an effort to minimize cleavage failure susceptibility.

(3) Shear

In a joint subjected to shear stress the force acts parallel to the plane of the adhesive in a manner tending to slide

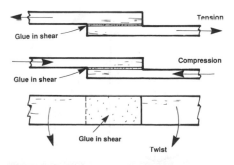

Tension

Glue in shear

Compression

Glue in shear

Glue in shear

Twist

Single lap joints.

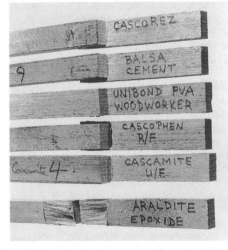

Wooden test pieces glued with various adhesives. Lower specimen shows fracture entirely in the wood.

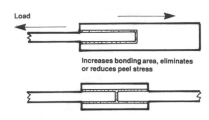

Double lap joints.

arrangement where the force on the two parts may be either pulling them apart or pushing them together, the effect on the adhesive is the same in each case. Additionally the force applied could be tending to twist the pieces out of line in the plane of the joint.

Adhesives in shear produce strong joints. It is usually the preferred design technique for accommodating tension loads since it is possible to largely avoid

one surface over the other, the function of the glue is then to prevent this sliding.

The common lap joint represents the simplest configuration of such an

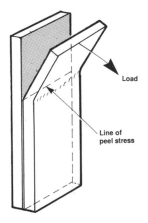

Peel failure..

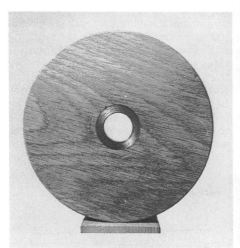

A concentric joint. Steel boss glued into wooden pulley using Araldite epoxide adhesive.

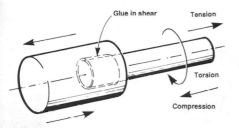

Concentric joint.

or overcome the risk of peel occurring. Single lap joints can be used in many instances but the double lap joint offers much greater strength and reliability where it can be incorporated. The latter may be built up from separate components although in critical assemblies machined grooves provide maximum resistance to adverse forces.

An excellent variation of the shear joint is possible for assemblies having one or more cylindrical components which can be fitted into a corresponding cylindrical hole machined to a reasonably close fit — a gap of one or two thousandths of an inch is usually called for. Able to absorb tension and compression forces along the axis of the cylinder, it also deals equally well with torsional forces. Being cylindrical in form there is no beginning or end to the adhesive layer so peel is not a problem.

COMBINATION JOINTS

Familiar to woodworkers, the scarf joint consists of two parts meeting along a slanting face. Its object is to present a larger gluing area than would be obtained if the same pieces were merely butted together, the angle of the scarf determining the length of face and hence the area.

It is important to note that an added benefit results from the adhesive layer being able to absorb tensile loads by a combination of shear and tension rather than in tension only, as in a plain butt joint.

Other more complicated forms of scarf are possible, the object being to avoid the disadvantages of making each piece taper accurately right down to the point, and the fact that such a thin portion might become distorted in service with damage to the glue line bringing the danger of peel from that end. The splayed scarf eliminates the sharp point but is more difficult to shape to a close fit.

Another method of increasing the joint area of a tension joint without altering the overall dimensions is to groove the faces in various ways. Plain parallel grooves fitted with keys may be adopted or a tongue and groove system. Single or multiple Vee grooves gives an effect similar to that of the scarf but doubled back on itself. A one-piece multigrooved cutter is almost essential to be certain of a close fit right across the joint face when adopting this method.

Other techniques that may be borrowed from the woodworker for increasing the strength of glued joints in metal or plastic structures are concerned with the provision of 'loose pieces', that is to say additional pieces of material that add extra area to the glue line for reinforcement. Triangular corner blocks, strips to thicken edges, keys across mitred angles and dowels all fall into this category.

When considering the strength of a glued joint both the nature of the material to be joined as well as the adhesive itself need to be taken into account. For example, wood adhesives are often quoted as having a strength greater than 200 kgf/cm^2 (2800 pounds per

Scarf joint repair to old chair back. Mahogany glued with urea/formaldehyde adhesive (before staining and finishing).

square inch), indicating that the wood itself is likely to fracture before the glue fails. Many adhesives including the old animal glue are spoken of as being stronger than the wood since at failure the glue line is often largely intact although breakage may have occurred very close and parallel to it. For this reason such a figure should be used in calculations even when the adhesive is capable of forming stronger bonds with other substrates. The strength refers to lap joints parallel to the grain of the wood, butt joints in end grain is

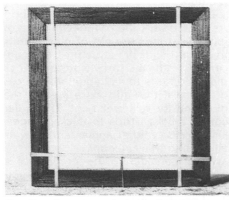

Small mahogany and glass window for instrument. Elastic bands hold grooved strips to glass while the PVA glue sets in the mitre joints.

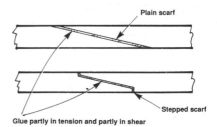

Glue partly in tension and partly in shear

Scarf joints.

Providing larger glue area.

a very much weaker configuration not normally employed.

Shear strength figures normally given are measured on small test pieces usually about one centimetre square. If the overlap is kept the same but the width is doubled, we would naturally expect the joint to be twice as strong, or if the width is three times as great to be three times stronger, etc. In other words the strength is proportional to the width of the joint.

Similarly we might expect the joint to be twice as strong if the overlap were doubled and so on, but in practice it is found that stress in the adhesive tends to be concentrated at the ends of the joint. This means that adhesive at the centre of the joint may be carrying much less of the load so that strength is not directly proportional to length of overlap. In simple terms it can be said that somewhat more than twice the overlap should be provided if the joint is to carry double the load, but it is not

easy to calculate exactly unless trials are made with the particular substrates and adhesive under consideration.

We see then that the load carrying ability of a joint may be increased by either increasing the width of the joint or the length of overlap or both. It will be appreciated that with metal substrates, failure in the metal is only likely for very thin strips since metals are so much stronger than adhesives. Even the strongest adhesives will fail before the metal if the latter is more than about one and a half millimetres thick in the case of aluminium alloy, and of course less than this for steel.

Overlaps in shear joints should not normally be less than about one centimetre if maximum strength is to be achieved, but this will obviously be dependent upon the job in hand — for scale models smaller overlaps may be a necessity.

The strength of an individual adhesive is often quoted in N/m² which means Newtons per square metre; this is a very large unit so the more convenient N/cm² is sometimes seen. MN/m² or kN/m² meaning mega or kilo Newtons refers respectively to millions or thousands of Newtons.

To put this into more familiar units,

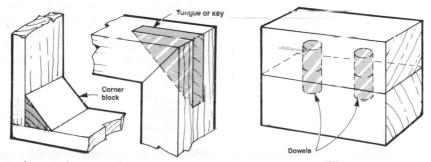

Loose pieces to increase glue area.

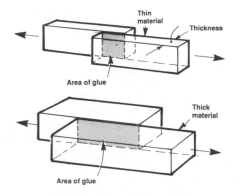

Area of glue related to thickness of material.

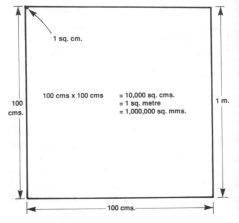

Metric units of area.

one kilogram force is roughly equal to 10 Newtons or for a more exact figure 9.81 Newtons; therefore the strength of an adhesive given as 500 Newtons/m² is the same as 50 kgf/m². In other words, an area of one square metre of adhesive surface would support 50 kilograms before breaking. One MN/m² would imply that one square metre of adhesive surface would support 100,000 kilograms before failure.

There are 10,000 square centimetres in one square metre which means that one MN/m² is the same as 100 N/cm² or 10 kgf/cm² and this latter measurement is rather more meaningful to most of us, especially for the small areas of glued joints we are likely to use.

It should be remembered that even for one particular adhesive product, its strength will be different for each type of material it is used to join. Care must be taken to apply to any calculations the strength stated for the combination of adhesive and material chosen for the job. Should a figure not be given it will be necessary to carry out a test to determine the strength before the design can be finalised. If a figure for the strength of a joint in mild steel is given it must be understood that the same figure does *not* apply to a joint made in brass with the same adhesive, nor with any other metal, because the surface adhesion will be different for each. Other values again would be obtained for wood, ceramics, glass, plastics etc.

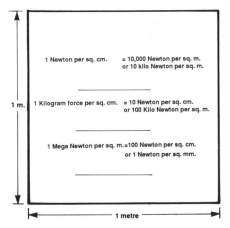

Units of adhesive strength.

It must be emphasised that each type of plastic will have its own joint strength in combination with each type of adhesive – there is a very wide difference in behaviour in this group. Beware of claims that a certain adhesive product is suitable for plastics, what is meant is that it may be suitable for some plastics. The responsible manufacturer lists those it will stick and often indicates those for which it is not recommended.

Sometimes a strength figure for mild steel is quoted together with the information that the same adhesive, when used on brass, will have a strength equal to the figure for steel multiplied by a given factor. For example, the bond strength when two pieces of mild steel are joined might be 10 MN/m^2 and the strength factor for a similar joint in brass be 0.7, therefore 10 MN/m^2 × 0.7 = 7 MN/m^2. For zinc it might be 0.3 and 10MN/m^2 × 0.3 = 3 MN/m^2.

Look carefully to see what sort of test the strength refers to, the most usual is tensile shear strength, but direct tension strength or compression strength might be quoted in some cases where it is relevant to the use of the product. Each can only be used in calculations where that type of stress is dominant.

As mentioned earlier one form of failure possible with both direct tension joints and lapped shear joints can be by peeling of the adhesive from the substrate. Once initiated this might progress rapidly in zip fastener fashion, particularly under rapidly changing or shock loads.

Adhesive data sheets often include *peel strength* values. As would be expected these vary considerably according to the adhesive/adherend chosen and the problem seems to be pronounced with very smooth surfaces such as those found on glass, glazed ceramics, polished metals etc. By removing the 'shine' from susceptible surfaces by abrasion or chemical methods, an improvement in peel strength can be achieved but it is still a potential source of weakness that has to be guarded against or designed-out of the system.

Since peel starts from one end of a joint then works progressively along the length of the bond it will be evident that at any given time during the failure separation is taking place on a single line across the bond. As a line has no area the stresses acting on the adhesive undergoing failure are very high although the actual force causing the failure may be quite low. This is an example of stress concentration comparable to crack propagation in a notched metal specimen.

Peel strengths listed in adhesive data sheets are given as so many Newtons per centimetre or Newtons per millimetre, i.e. N/cm or N/mm. For our convenience we may change these into kgf/cm or kgf/mm merely by dividing the number by ten. In some cases mixed

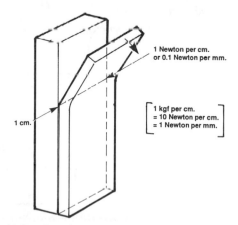

1 Newton per cm.
or 0.1 Newton per mm.

1 kgf per cm.
= 10 Newton per cm.
= 1 Newton per mm.

1 cm.

Units of peel strength.

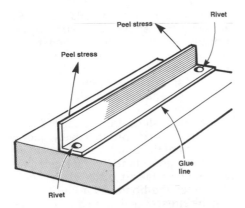

Rivet at end to prevent peel.

pull the strips apart from one end will be $5 \times 2 = 10$ kg. The total length of the bond is immaterial, it is only its width that matters, so a force of 10 kg would just keep on tearing it apart right to the other end.

When adhesives first began to be used in large structures it was initially thought that all mechanical fastenings in the form of rivets, screws, bolts, etc., might be eliminated. It was soon found though, in aircraft wings for instance, that the fluctuating loadings induced by turbulence, vibration and the shock of landing led to peel separation at critical points. It is now considered good practice to employ rivets in those areas where peel might be expected to occur. In these cases the whole seam is glued and in addition some rivets used, especially at the ends of the seam.

units are given, for instance kg/inch, but this should be readily interpreted. Note that the cm, mm, or inch refers to the length of the line along which the adhesive is breaking away, for example if two 5 cm wide strips are glued together with an adhesive having a peel strength of 2 kg/cm, the force needed to

Incidentally some improvement in peel resistance can be achieved in certain types of joint by allowing a relatively thick layer of adhesive. This

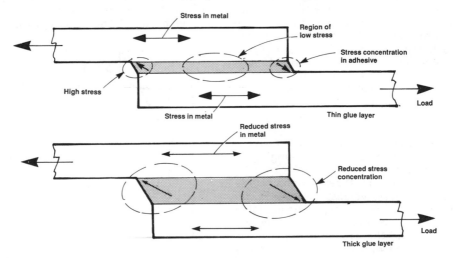

Effect of adhesive thickness.

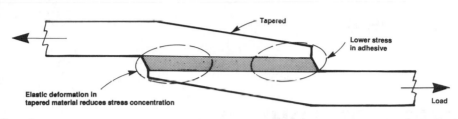

Elastic deformation in
tapered material reduces stress concentration

Tapered

Lower stress
in adhesive

Load

Tapering.

enables the adhesive to deform under load without breaking and thus transfers the stress over a greater joint area. It is more pronounced with the newer toughened adhesives which have greater flexibility. Usually one aims to ensure thin glue lines so this method represents an exception to the accepted rule.

A similar effect is gained by tapering one of the components at the end of the joint. As with the scarf joint this reduces stress concentrations just where the danger is greatest by absorbing energy in elastic deformation of the substrate.

SUMMARY OF DESIGN CONSIDERATIONS
By being aware of the strengths and limitations of adhesive systems, knowing something of the nature of the adhesive bond together with the manner in which stresses may best be resisted, it should be possible to arrive at a satisfactory joint design to suit most of the situations we are likely to encounter. The points to consider are:

(1) The type of materials to be joined, i.e. wood to wood, steel to steel, brass to brass, rubber to rubber, etc, or wood to steel, rubber to brass, and so on. In cases where dissimilar materials are to be bonded the peculiarities of each must be allowed for.

(2) The type of adhesive to be used, i.e. it must be compatible with the substrates to produce bond strengths reasonable for the particular job. It must also be capable of withstanding the conditions of temperature, humidity, oil or other chemicals likely to be met in service. It must be in a form suitable for application to the surfaces, to fill the joint and remain in place until set.

(3) Sufficient joint area must be provided to accommodate anticipated stresses, including minor accidental and shock loadings. A factor of safety should be used to increase minimum joint area up to a size adequate for all reasonable eventualities.

(4) In deciding the joint area desired the best configuration to resist the expected forces must be found. Where possible arrange for the adhesive to accept the load in compression or shear.

(5) Avoid abrupt changes of section at the ends or edges of joints where peel stresses are likely to be significant. Where possible avoid vulnerable points by redesign or addition of loose pieces to cover or reinforce such regions. Where this is not practicable consider strengthening the region by use of rivets or other mechanical restraints.

CHAPTER 4

TYPES OF ADHESIVE

There are a number of ways in which adhesives may be classified including Chemical Composition, Form Supplied In, Mode of Curing, and so on. Of these it is probably best to first gain an understanding of the manner in which they set since this will in many cases limit the jobs they may be suitable for.

SETTING BY EVAPORATION
In this group there are two distinct sub-groups that at first glance appear to be quite different but do work in basically the same fashion:

(a) Water-based adhesives
By their nature adhesive substances softened by water do not actually dissolve in it as sugar dissolves in water. What happens is that the substance absorbs water and swells up into a sticky, jelly-like mass known as a *colloid*. This sort of behaviour is found with animal glue, starch glue and the cellulose type used for hanging wallpaper. These are relatively safe to handle with no fire risk and low inherent toxicity, although added substances to prevent deterioration through mould or bacterial growth can be toxic if ingested.

Because the glue hardens by evapor-

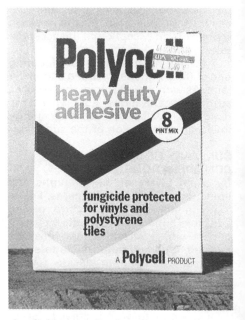

A colloid-type adhesive. Polycell methyl cellulose wallpaper adhesive powder for mixing with water.

ation of water there will be some shrinkage during setting and it means that this type is only suitable where at least one of the materials being joined is

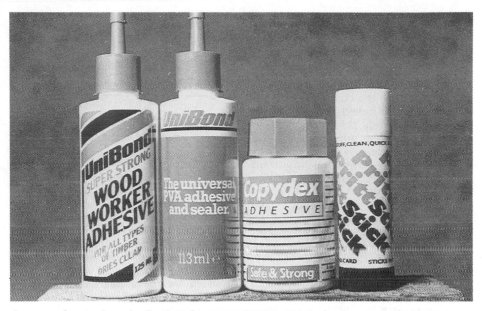

A range of water-based adhesives for paper, fabrics, wood, etc.

porous. With two non-porous surfaces evaporation could not take place. Materials like paper, wood, fabrics, leather, etc, first absorb the water from the glue line then allow it to permeate to the surface and evaporate.

This sort of adhesive is only suitable for use in dry conditions, it softens rapidly when allowed to become damp with consequent loss of bond strength. It must not be thought that this is always a disadvantage, just think how difficult it would be to remove old wallpaper otherwise.

There Is a modern tendency towards water-based adhesives of this and the following type in order to remove the 'glue sniffing' hazard as well as to eliminate loss of expensive solvents into the atmosphere which only adds to pollution problems.

Another form of water-based adhesive is the *emulsion* or *latex* which was mentioned in the first chapter. This is a means of providing water-based products in which the adhesive substance itself is not soluble in water but is dispersed in liquid droplets held in suspension in water. Examples include the natural rubber latex as well as PVA, acrylic emulsions and others.

When the water evaporates the polymer droplets mix together before drying as a thin solid film attached to the joint surfaces.

(b) Solvent-based adhesives

A solvent in this context refers to an organic chemical that exists as a liquid having the characteristic of readily evaporating at room temperature. It must also of course dissolve the adhesive substance either on its own or in combination with one or more other solvents. Mixtures of solvents are often used for reasons of obtaining improved solubility, to control the rate of evaporation or to reduce the cost.

As with water-based adhesives this solvent type is only able to function in a closed joint when one or both of the surfaces is porous.

There is however another important way in which they may be used with non-porous substrates. This is known as the *contact* or *impact* adhesive method. In this case a thin layer of the adhesive solution is spread evenly over each surface and the solvent allowed to evaporate for a certain length of time. At this point the adhesive layers may be tacky to the touch or, in some instances, may appear to be quite dry. When the two coated parts are placed in contact they immediately stick together, although the full strength of the joint is not developed until considerable pressure has been applied to fully unite the glue. This is sometimes achieved by

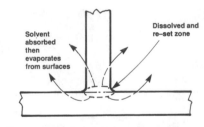

Effect of solvent on some polymers.

hammering the surfaces together with a rubber mallet to avoid damaging the finish, hence the term *impact*.

Solvent-based adhesives generally have good resistance to water although they are softened by chemicals similar to the solvents originally used and can be affected by oils and greases. They tend to be more flexible than other types particularly the contact ones.

HOT MELT ADHESIVES
These modern versions of the old shellac and sealing wax adhesives are synthetic polymers exhibiting good qualities of adhesion while in the mol-

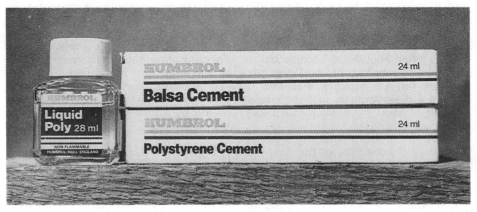

Solvent-type adhesives for balsawood and polystyrene.

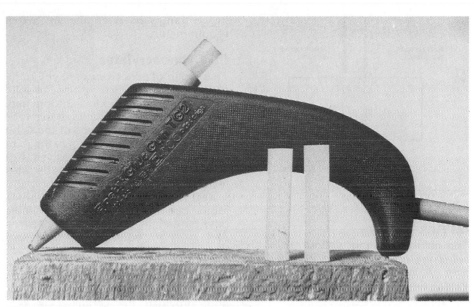

Electric hot-melt glue gun with sticks of adhesive.

ten state; upon cooling they revert to a solid form within the joint.

Being thermoplastic polymers they may be heated and softened again and again. Supplied as short pencil-shaped sticks they are intended to be applied from electrically heated 'glue guns'. Only the required amount is squirted from the gun; any surplus is left inside to cool and be reused later. Since the joint must be closed before the melt can appreciably cool it is only really suitable for small fabrications, the strength of bonds is somewhat limited.

ONE-PART POLYMERISATION ADHESIVES

This category too may be divided into two parts: those that set in the absence of air and those that set by reaction with water vapour from the air.

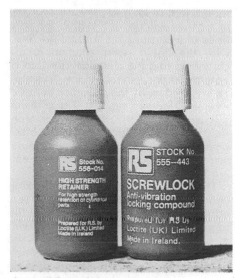

Anaerobic retaining compounds.

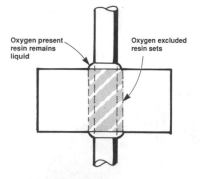

Anaerobic adhesive.

(a) Anaerobics

The name, meaning without air, indicates the manner in which this class sets. A number of synthetic resins exhibit the peculiarity of being reluctant to polymerise in the presence of oxygen; they are said to be oxygen inhibited. Since the atmosphere contains a large proportion of oxygen, this means that as long as they are exposed to air they will not set. Within the confines of a closed joint completely filled with the liquid resin, oxygen is excluded so that the chemical reaction of polymerisation can take place. This may be speeded up by the presence of metal oxides acting as a catalyst. These resins are usually versions of the acrylic type. Being thermoplastic they will soften when heated but regain the original strength on cooling provided the heating has not been too drastic.

The types used as threadlocking compounds generally have little real adhesion to the metal but work by filling the thread clearance spaces with a solid resin film that increases the friction. By altering the composition of the resin it is possible to provide different frictional characteristics mak-

ing a range of strengths to match requirements.

(b) Cyanoacrylates

This group of adhesives, containing the chemical cyano-structure, also set solid when confined in a close fitting gap. Polymerisation of the liquid in this type is initiated by the presence of the layer of water molecules adhering to the joint surfaces. These are true adhesives with an affinity for a number of substrates including some plastics and rubbers.

Best known by the speed with which joints can be made, cyanoacrylates have earned the title 'instant' adhesives and care in handling is necessary because of the ability to bond skin tissue. Although expensive it is an economical way of bonding small parts where often a single drop will suffice and, provided the container is kept tightly sealed against the entry of atmospheric moisture, the contents will remain usable for some months at least.

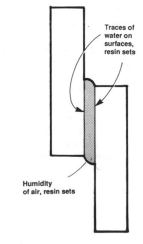

Instant adhesive.

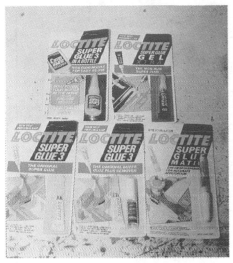

Some cyanoacrylate 'instant adhesive' variations. (Photo courtesy Ted Jolliffe).

TWO-PART POLYMERISATION ADHESIVES

This type usually consists of a resin and a hardener both of which are reactive chemical compounds in a semi-liquid state. When mixed intimately together irreversible chemical changes take place whereby the mixture sets to a solid mass. This form of polymerisation results in the formation of giant molecules that do not become plastic when repeatedly heated; a thermoset resin like this undergoes a once-and-for-all setting process.

With the epoxide resins the hardness and flexibility of the cured mass depends upon the chemical structure of the hardener used. A number of different hardeners are available giving not only altered physical properties but also setting times from a few minutes to several hours. When polyamide hardeners are mixed with epoxide resins the proportions may be varied either side of the 50/50 level; less hardener making the set resin more rigid, while more hardener produces a more flexible product. In this sense the word 'hardener' would appear not to be a good descriptive choice.

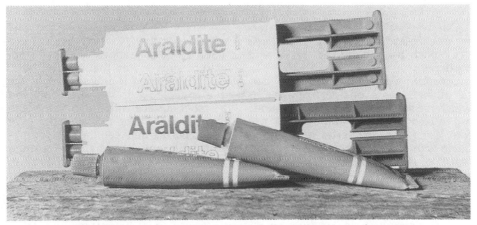

Two-part polymerisation-type adhesive. Squeeze tubes and syringes of epoxide resin/hardener.

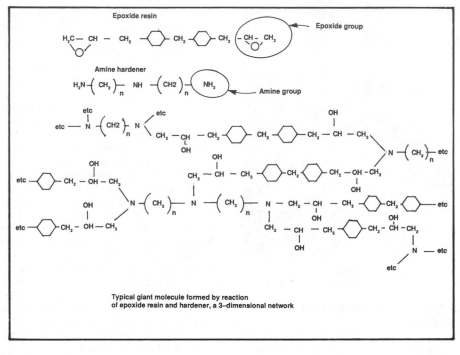

Epoxide adhesive chemical structure.

Other adhesives such as phenol/formaldehyde and resorcinol/formaldehyde use fixed proportions of resin and hardener.

Urea/formaldehyde adhesives can be obtained as a one-part powder in which the solid resin and solid hardener are mixed together. Immediately before using, the powder is mixed to a paste with water thereby activating the hardener and starting the reaction. It is essential to keep this powder adhesive in an airtight container as atmospheric water vapour is sufficient to cause slow reaction during storage. At first this would mean a loss of strength, later it would become lumpy when mixed and eventually the powder would set solid in the tin or jar.

Some two-part adhesives are formulated in such a way that a good bond may be obtained if one surface of the joint is coated with the resin only and the other surface with a thin film of the hardener only. Upon closing the joint mixing occurs between the two components, but until this happens setting cannot commence. There is plenty of time to cover large or intricate surfaces or to make sure of accurate alignment. Another advantage is that the exact amount of each is more easily gaugéd thus avoiding wastage. With pre-mixing one tends to overestimate the quanti-

Isopon car body filler – a two-part unsaturated polyester adhesive.

ties and any adhesive mixed but not used will not keep so must be thrown away.

The well known polyester resins used on a large scale for the manufacture of fibreglass mouldings can also be used as an adhesive but set in a slightly different manner. The liquid polyester comes already mixed with the liquid monomer styrene which gives the uncured resin its distinctive odour. Another ingredient is supplied in a separate small container, the catalyst. It is mixed into the resin just prior to using, which has the effect of starting a chemical reaction between the polyester and the styrene during which the long chain polyester molecules become joined to each other at intervals along their length. This 'cross linking' results in the formation of a complex three dimensional network or giant molecule. This is a thermoset compound that cannot be reshaped by heating, although it will become softened at high temperatures. It cannot be dissolved by solvents either.

These then are the main categories into which adhesives may be grouped.

Within each group there can be a diverse range of products each designed to adhere to certain materials in certain types of joint.

We can now attempt to list various adhesives within these sections:

(1) Water-based colloids

Animal glue (gelatin), fish glue (Isinglass), starch glue and its derivative Dextrin glue, Methyl Cellulose (Polycell), Polyvinyl Alcohol.

(2) Water-based emulsions

Rubber latex, Polyvinyl Acetate (PVA), Casein, Acrylic.

(3) Organic solvent-based

Rubber solution, Polyvinyl Acetate (PVA) solution, Polystyrene solution, Acrylic Ester solution, Cellulose Acetate solution, Polyurethane solution, Phenolic Resin solution.

(4) Hot Melt/Set on Cooling

Shellac, Waxes, Bitumen, Various Thermoplastic Polymers.

(5) Anaerobics, Set in Absence of Air

Certain Acrylic Ester and Polyester types.

(6) Instant, Set in Presence of Water

Cyanoacrylates.

Also slow setting types:

Silicone sealant, certain polyurethanes.

(7) Two-part Polymerisation.

Pheno/Formaldehyde, Urea/Formaldehyde, Resorcinol/Formaldehyde, Epoxide/Amine, Epoxide/Amide, Polyurethane, Toughened Acrylic, Polyester.

Further classification may be attempted by considering which adhesives are suitable for bonding individual materials, but before looking at that aspect we need to examine those materials in some detail. The properties of materials differ widely so both their chemical constitution and their physical form affect the manner in which they may be bonded together.

SOME TYPICAL INDUSTRIAL USES OF ADHESIVES

Headlights for the Ford Escort being bonded with Araldite by Lucas. (Photo courtesy Ciba-Geigy Plastics)

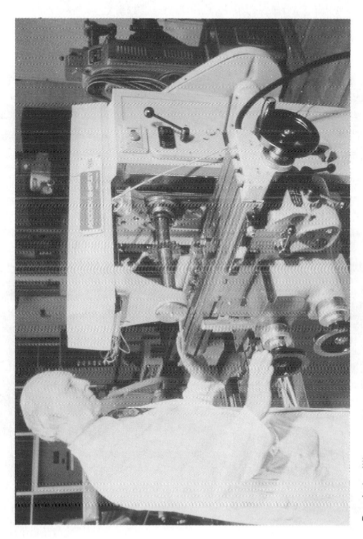

Bonded milling machine. (Photo courtesy Ciba-Geigy Plastics)

Brake bonding process at Automotive Products plc, Leamington Spa. Random brake samples tested to destruction.
(Photo courtesy Ciba-Geigy Plastics; photo by: Michael Manni Photographic)

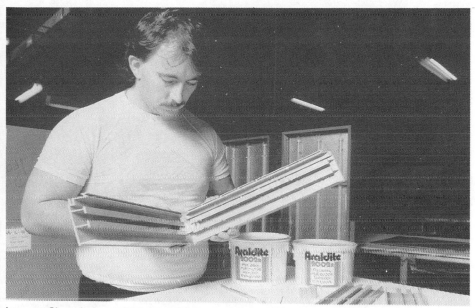

Lynesta Signs Ltd use Araldite 2002 to construct a wide range of illuminated signs.
(Photo courtesy Ciba-Geigy Plastics; photo by: Michael Manni Photographic)

The Maestro is assembled with the aid of Araldite adhesive by Austin Rover.
(Photo courtesy Ciba-Geigy Plastics; photo by: Reeve Photography)

Sandwich bonding at York Thermostar. Araldite system for bonding metal to wood to PVC to metal in production of side and roof panels for refrigerated container trailers. Metal outer sheet being rolled onto PVC to finish the bonded sandwich sheet.
(Photo courtesy Ciba-Geigy Plastics; photo by: Michael Manni Photographic)

Finished refrigerated container trailer ready for deliverage.
(Photo courtesy Ciba-Geigy Plastics; photo by: Michael Manni Photographic)

50

CHAPTER 5

THE MATERIALS TO BE BONDED

From a very early age we learn to stick things together, the material most often involved is probably paper in various forms including card and cardboard. These are all made of cellulose which is the main constituent of wood; cotton, too, is practically pure cellulose. When wood is broken down into pulp for papermaking the individual wood fibres retain their identity and are merely separated from each other. The lignin which bound the fibres together in the tree is removed by chemical treatment before the pulp slurry is allowed to settle out on wire mesh frames. During the settling process the flexible wet fibres become entangled with each other or 'felted together' so that when dried a continuous sheet is formed. When paper is torn many of the fibres are simply pulled loose from the sheet although some may be actually broken. A good magnifying glass or low power microscope will clearly show the fibrous structure, particularly at a torn edge.

Paper is a good example of a porous adherend — we know that when it is wetted the water is soaked up as by a wick. A water-based adhesive too will be drawn right in among the fibres enabling the set glue to be mechanically interlocked as well as adhering to the fibre surfaces. Organic solvent glues can penetrate the paper in the same way to obtain a secure grip.

Paper used to print glossy magazines is made by mixing the pulp with fine china clay powder and other fillers to give a dense white surface, then various sizing agents are added to bind it all together. In this way the gluing characteristics are altered making a strong bond harder to obtain though the material is still porous.

When sheets of paper are impregnated with phenolic resin, stacked together and subjected to heat and pressure, a solid hard sheet of brown 'plastic'

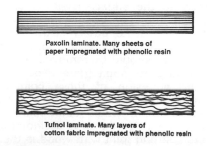

Paxolin laminate. Many sheets of paper impregnated with phenolic resin

Tufnol laminate. Many layers of cotton fabric impregnated with phenolic resin

Paxolin and tufnol.

51

material known as Paxolin is formed which is widely used in the electrical industry as an insulating medium, for making printed circuit boards and cabinets. The paxolin is a non-porous material because the resin has completely filled all the spaces between the fibres as well as thoroughly coating the surface of each fibre. When an adhesive is applied to the surface it cannot reach the cellulose component, therefore it is in effect being used to bond phenolic resin which is the same as Bakelite. The type of adhesive must then be one that will stick to a waterproof impervious thermoset plastic rather than a paper adhesive.

Tufnol is a very similar product, also having layers of cellulose reinforcement in a phenolic resin, but this time it is woven cotton cloth that makes up the layers. The gluing characteristics however are the same as for paxolin.

The need often arises to glue together pieces of wood, which we have seen is predominantly cellulose, with the fibres aligned in one direction to form the 'grain' structure. Wood is a porous material capable of absorbing water and other solvents. Certain adhesives having an affinity for cellulose can penetrate a short distance into the wood below the joint surface to secure an interlocking bond, others may just adhere to the surface itself. When a wooden item has been painted, varnished, polished or otherwise treated, it is important to appreciate that such substances should be removed before attempting to apply glue. Not only is the adhesion of these coatings usually rather poor, so that a bond made to them would soon fail, but also an adhesive that was designed to make a strong joint with wood is then not actually in contact with cellulose but probably with a 'plastic' or waxy material with which it is unlikely to be compatible.

Not all varieties of wood respond equally well to adhesives. In general the softwoods, like spruce, larch and pine, take glue more readily than the hardwoods, such as elm, ash, oak, beech, birch, mahogany, etc. Excessive amounts of resin in the softwoods can create problems which are best dealt with by rejecting those particular pieces where visual inspection or lack of wetting show that resin is obviously present. Beech and birch often present a rather shiny surface when freshly planed which should be removed with a medium grade abrasive paper at the joint surfaces before application of the adhesive. A particular problem is encountered with teak because of its oily nature, but rubbing over the joint area with a clean cloth soaked in suitable oil-removing solvent a few times should allow a sound bond to be made.

Wood is capable of absorbing moisture from the atmosphere as well as from damp surroundings. The effect of this is to swell the wood across the grain so that its dimensions are changed from the dry state. If glued with too high a moisture content the bond may well be weakened because the adhesive is unable to soak in properly, and on drying out the joint might be under considerable stress as the material shrinks again. It is wise to store timber for several days in the workshop prior to final shaping and gluing.

Another group of porous materials is the fabrics or woven fibres. The name fabrics is a very loose description which only indicates the physical structure of the product. Within this group can be a wide range of chemically different materials.

The natural and long established members will be recognised as cotton, silk, wool, linen and flax. Then there are intermediates made by chemical modification of natural fibres in the form of viscose (cellulose) and cellulose acetate followed by the purely synthetic fibres nylon (polyamide), terylene (polyester) and acrylic (acrylonitrile).

Those based on cellulose form a large proportion of this total, being cotton, linen, flax, viscose and cellulose acetate. The latter differs from the others in that it will dissolve in some organic solvents, for instance acetone, which explains why acetate clothes disintegrate when nail varnish remover is spilt on them. A solution of cellulose acetate in acetone is an adhesive that can be used to stick solid cellulose acetate sheet but would damage a cellulose acetate fabric by dissolving it before the solvent could evaporate.

Silk and wool are both animal protein fibres so their chemical structure resembles that of nylon. They smell similar when burned due to traces of ammonia in the smoke. We should therefore expect them to have the same sort of affinity to certain adhesives.

Polyacrylonitrile molecules contain nitrogen as do the previous group, but in a slightly different chemical structure, and it has some resemblances to them in its properties too.

Terylene is a polyester having the property of being capable of drawing out into long fibres. Its proper chemical name is *polyethylene terephthalate*.

Both polythene (polyethylene) and polypropylene are made into twines that appear on casual inspection to be formed from fibres but closer observation reveals them to consist of slit and rolled films. There is in fact a limited number of plastics that can be drawn

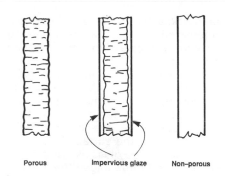

Porous   Impervious glaze   Non–porous

Ceramic types.

into fine strong fibres suitable for spinning and weaving into fabrics.

Still on the subject of porous materials we come to a group often referred to as ceramics, though some of these are non-porous in nature either on the surface only or throughout their mass.

The porous ones include most types of brick, certain tiles and pottery and some sorts of china. They are made from clay fired in a furnace to a temperature sufficient to fuse together the individual grains only at the points where they touch each other. The irregular shape of the grains means that there are spaces between them forming a sponge like network of narrow passages right through the apparently solid shape. This feature can be demonstrated by placing a piece in water and noting its absorption. Water- and solvent-based, as well as pure liquid, adhesives can soak into the material along the joint line to form an interlocked layer.

Those ceramics that have been fired at a higher temperature fuse completely into a solid impervious mass having a glass-like surface and showing the same glassy fracture when broken. Only those adhesives capable of forming a chemical bond with these surfaces can be expected to make a strong joint with this sort of material. Items belonging to this group are labelled stoneware or porcelain.

Glazed ceramics are those consisting of a porous body which has received a thin coating of slurry that is allowed to dry before firing for a second time. During this firing the body remains unchanged but the surface layer fuses into a continuous glass-like film. When complete and undamaged the article is completely waterproof, but once cracked or broken water can be absorbed. Repair is easier than for the porcelain type but the broken surfaces must be dry and clean for the adhesive to penetrate.

Glass itself is a difficult material to bond partly because of its extremely smooth shiny surface and complete impermeability, but also on account of its chemical structure. Water has a strong affinity for glass and in exposed situations will significantly weaken adhesive bonds.

SOLID POLYMERS (PLASTICS)
There is a very wide range of chemical types within this group of non-porous materials. As with some of the synthetic adhesives mentioned in the previous chapter a division into thermoplastics and thermosetting plastics is advantageous.

Thermoplastics
One of the earliest of these to be developed was cellulose nitrate. This substance had been in use for some time as a smokeless replacement for gunpowder and was used on a large scale in the First World War. In peacetime there were large stocks to be disposed of and it was also deemed

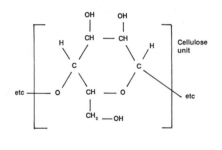

Cellulose unit

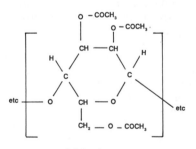

Cellulose Acetate unit

Many units joined together form one long chain polymer molecule

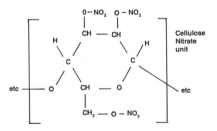

Cellulose Nitrate unit

Cellulose and its derivatives.

54

The 'phantom cricketer' model by an 11-year-old boy shows synthetic 'clay' pieces glued to wires and base with epoxide adhesive.

advisable to retain the manufacturing capability in case of future need. It had been found that by mixing a quantity of camphor with the cellulose nitrate (often incorrectly called nitrocellulose) a clear transparent, slightly flexible sheet material could be made. This celluloid was a very useful product but had two rather severe drawbacks. In the first place it was extremely inflammable and secondly it had a tendency to deteriorate with age becoming yellow, brittle and liable to spontaneous combustion. When dissolved in acetone or other solvents cellulose nitrate could be used as a paint or varnish, or indeed as an adhesive which was the original form of balsa cement.

Fire danger from cellulose nitrate cine film prompted the development of a non-flam alternative resulting in the introduction of cellulose acetate which is still with us. This material can be obtained in a wide range of colours as well as clear sheet. It too is soluble in some organic solvents like acetone. Plastics of this type that form an adhesive solution can be virtually welded together since the solution softens the substrate and amalgamates with it.

A similar plastic is polyvinyl acetate (PVA), again readily soluble in organic solvents and therefore easy to bond. This is a vinyl ester compound which has totally different properties to that other well-known vinyl compound, polyvinyl chloride or PVC.

PVC in its pure form is a hard material

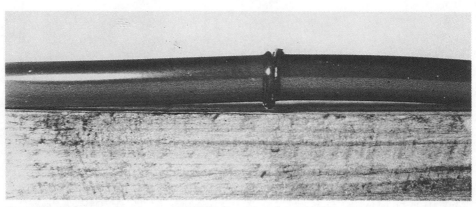

Round polyurethane belt heat-welded (before trimming joint).

with a tendency to brittleness, particularly following exposure to sunlight or other sources of ultraviolet (UV) radiation. In fact most plastics suffer from this tendency despite the incorporation of UV stabilising chemicals. Few organic solvents will affect PVC which is often regarded as being difficult to bond although there are special adhesives made for the purpose. In industry much use is made of plastic welding techniques which rely on heating the material and pressing the hot surfaces together so that they fuse into a continuous joint. A variation

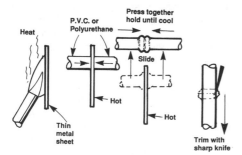

Heat welding plastic belting.

is to blow hot air or inert gas onto the joint surfaces often in conjunction with a filler rod of the same material as the substrate to build up the joint fillet. It is necessary to support these plastics during welding since the heat can soften the pieces causing sagging or distortion.

Small items may be welded together by simultaneously pressing both joint surfaces against either side of a thin metal strip heated to the welding temperature. Once softened the parts are quickly slid off the edge of the strip and immediately pressed together, being held in alignment until cool. An example is the making of small machine belts from round section PVC or polyurethane flexible cord.

The flexible form of PVC is obtained by mixing into the polymer another chemical compound called a plasticiser. In the early days these were simply organic liquids of fairly high boiling point, so that they did not evaporate, which swelled and internally lubricated the plastic. Their disadvantage was the 'leaking out' or migration of plasticiser to the surface where any adhesive

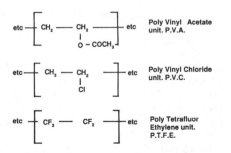

Typical thermoplastic molecules.

would be floated off. Modern plasticisers are more like polymers themselves so tend to stay put within the mass of the plastic. As with all plastics it is a good plan to wipe over the joint surfaces prior to bonding with a solvent-soaked clean cloth to remove excess plasticiser or other contaminant.

Polyvinylidenechloride is another chlorine containing plastic that can be produced in both sheet and fibre forms, though it is not made in anything like the same tonnage as PVC. It has more or less the same property of chemical inertness leading to bonding problems unless the correct adhesive is chosen.

Fluorine in the polymer molecule results in even more marked resistance to chemical attack. With polytetrafluorethylene the long chain molecules have a backbone of linked carbon atoms to which are attached the fluorine atoms, there being no hydrogen atoms present at all. PTFE is an extremely inert material with a 'slippery' feel to it and has a high softening temperature close to its decomposition temperature making it difficult to mould into shape. Great caution should be exercised when making parts from PTFE because when heated it breaks down to give very poisonous fumes. This is probably the most difficult plastic with which to make adhesive bonds. It can be done in industry by using special surface activating chemicals but due to hazards in their use they are not generally available. It is probably best for the amateur to avoid adhesive joints when making parts from PTFE and stick to mechanical fastening methods, but see page 107.

Another fluorine containing polymer of rather more complex chemical structure is found in one of the inert synthetic rubbers called a fluoro-elastomer, which is mostly found in the form of seals and 'O' rings.

Two similar polymers containing only carbon and hydrogen are the commonplace polyethylene and polypropylene. It is an interesting chemical fact that these two belong to a series of hydrocarbons whose physical properties vary from gases through liquids of increasing viscosity to waxes and finally hard solids. The gases butane and propane form one end of the series, petrol is a mixture of the more volatile liquids, lubricating oil a mixture of the thicker members, paraffin wax some of the

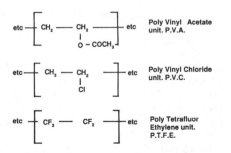

Similar chemical structures.

softer solids, and polythene and polypropylene the harder solids.

The only difference between these substances is the number of carbon atoms in each molecule. Propane has 3 carbon atoms in its molecule and butane 4, petrol has molecules with from 5 to 8 carbon atoms long, oil has about 12, paraffin around 16, but the polymers have thousands or tens of thousands of carbon atoms in each molecular chain. These substances all burn in a similar fashion; paraffin wax is used to make candles and polythene too will burn just like a candle.

This helps to explain why polythene is so difficult to bond; it has a waxy surface and just imagine trying to glue anything to a candle! We have here a clue though, since we know how easy it is to join two pieces of wax . . . melt the surfaces and press them together. The same can be done with polythene as in the heat-sealing of bags or thin sheets and the hot-air welding of thicker parts. We have also seen in Chapter 2 how the surface of polythene can be chemically modified to accept adhesives.

Polystyrene is a transparent hard plastic with a tendency to brittleness which can lead to cracking under stress. Coloured versions are also available as with other types of polymer. Toughened or 'high impact' grades have been developed to overcome the cracking problem — more will be said about these later. Polystyrene is an easy material to join using suitable solvent-based adhesives capable of softening the surface.

An excellent transparent plastic (though coloured forms are made) is that known as perspex or plexiglass, familiar for its use in aircraft canopies which commenced in the 1930s. Its strength as well as its resistance to weathering, including UV exposure, make it suitable for many applications. Its chemical name is *polymethylmethacrylate*, in other words it is one of the acrylic family of polymers. One method of joining perspex which used to find favour was the 'solvent bonding' technique whereby the joint surfaces were brushed with an organic solvent like chloroform until sufficiently softened to fuse together under light pressure. An alternative was to coat the surfaces with a solution of perspex in chloroform to make the joint. As soon as the solvent had evaporated a rigid bond with a strength similar to that of the parent material was obtained. Nowadays a number of adhesives will do the job.

Other acrylic ester plastics behave rather like perspex so are reasonably easy to deal with.

Polycarbonate is one of the more recent strong transparent polymers which appeared in the 1960s. Its most familiar form is perhaps the police riot shields we see on television which gives an indication of its toughness. In its early days the material suffered from a peculiar defect resembling the stress corrosion cracking of metals in that a stressed strip wetted with alcohol would suddenly crack. Whether riot shields disintegrate during drunken brawls is not known although the problem may have been cured by now. Such a phenomenon is not unique to polycarbonate as early polythene washing-up bowls also cracked following contact with detergent — reduction of residual moulding stresses, together with improved polymer grades, was the answer to that one.

The 'crazing', or development of a network of fine surface cracks, of certain polymers, especially following exposure to solvent vapours, can arise

Plastic watch glass showing 'crazing' cracks as a result of moulding stresses and exposure to solvents

during some fabrication operations but is not a common problem today.

Polycarbonate can be bonded using a solvent-type adhesive.

Acetal polymer is a tough hard plastic used for making precision mouldings such as small gears where strength in thin sections and resistance to wear is important. The material has quite a 'greasy' feel making it difficult to bond. The best of the products that I have been able to test proved to be Isopon P38 car body filler. Although not sold as an adhesive it is based on an unsaturated polyester resin mixed with styrene that becomes cross linked following the addition of a catalyst, so it is really a filled polyester resin. Even epoxide resins do not form a strong bond to acetal.

Polyester resin itself in the form of fibreglass laminate is extensively used for all sorts of structures, and it is a relatively straightforward matter to bond it with a number of adhesives.

Nylon in its solid form, as used for bearings and other components, is not a particularly easy material to glue although several products will form acceptable bonds for many applications.

CO-POLYMERS AND MODIFIED POLYMERS

Not all thermoplastics are made from one sort of starting molecule; by polymerising together two different types they may both be incorporated into the same long chain molecule in either a regular or a random alternation. By changing the proportions an optimum set of properties, including hardness, flexibility, softening temperature, ease of moulding, etc, may be achieved. A good example of a co-polymer showing the superiority over its main constituent alone is a high impact polystyrene in which the inherent brittleness of ordinary polystyrene is markedly reduced by the incorporation of a small amount of acrylonitrile into the molecules.

A modified polymer is again represented by a high impact polystyrene, this time a small quantity of butadiene-styrene co-polymer is mechanically mixed with polystyrene to render it less brittle.

A mixture of three components makes the useful tough plastic known as ABS, standing for acrylonitrile-butadiene-styrene, capable of withstanding a lot of hard usage.

Even PVC can be changed to a more flexible material by co-polymerisation of vinyl chloride with vinyl acetate. A small amount of the latter, about 15 per cent, being sufficient to alter the rigid structure of straight PVC to a more energy-absorbing form. A further gain is that the product is soluble in organic solvents like MEK (MethylEthylKetone) with which it forms an adhesive solution.

It is probably true to say that co-polymers and polymer mixtures present less of a bonding problem than some straight polymers.

THERMOSET POLYMERS

Bakelite, made by the reaction of phenol with formaldehyde, was one of the first plastics to be commercially manufactured. The product was a very dark hard brittle material that in its initial state had been capable of being shaped under pressure in a mould. Once heated to complete the reaction it hardened into its final form after which it could not be re-softened by heat. Since articles were usually made in highly polished metal moulds the surface of the plastic is itself very smooth and shiny but when broken the interior reveals a rough jagged fracture surface.

When attempting to repair broken items there is usually no problem in

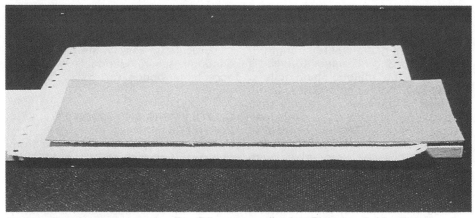

Sticking Formica to wood using Evo-Stik contact adhesive. Right-hand side aligned and lightly stuck ready to slide paper sheet away to avoid trapping air in joint.

Fibreglass test piece glued with resorcinol/formaldehyde adhesive. Fracture shows exposure of glass fibres.

securing adhesion to the broken surfaces although it is necessary to abrade the shiny exterior face of any joint if that also has to be glued.

These phenol/formaldehyde plastics are still around due to their cheapness and good electrical and mechanical properties. They have been joined over the years by other similar materials of which urea/formaldehyde and melamine-formaldehyde must be the best known. The former, for instance, may be seen in the white or light coloured electrical switches and fittings, the latter as hard plastic tableware and decorative laminates of the Formica type. There is little difference in the bonding behaviour of the members of this group.

The related tufnol and paxolin have been mentioned earlier in this chapter, the components of which are cotton or

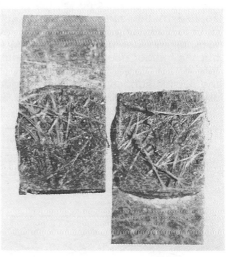

Fibreglass strip joint made with superglue. Fracture within the material exposes the glass fibres.

wood fibres laminated with phenol/formaldehyde resin. They should therefore be treated as Bakelite where their gluing characteristics are concerned.

Polyester resins may be found as thermoplastics when used on their own (see terylene under fibres), or they may be used as one component of a thermoset plastic. The material often referred to as fibreglass usually consists of layers of very fine glass fibres bonded together into a solid sheet with a thermoset polyester resin. Although frequently made in this fashion as a laminate it is also possible to mould articles from a mixture of short chopped glass fibres dispersed in the resin to give a dough-like consistency – the so called dough moulded products. The materials go under another name too, as 'composites'. They may be regarded as bonded layers or as reinforced resins with the fibres providing the tensile strength. Special 'gel coats' applied to the surface of the mould prior to laminating give a smooth shiny face to the exterior of the mouldings, but the interior usually shows the pattern of the fibres. To repair or fix to the outside surface it is best to abrade away the gel coat until the first layer of glass fibres is exposed. Since they are embedded in polyester resin it is essentially a question of choosing an adhesive that will stick to the latter.

While this does not exhaust the list of polymeric materials it does cover sufficient of those in common use to give a reasonable idea of the variation. There is however yet another group, not fundamentally different in chemical structure but whose physical properties are significantly changed. These are collectively called the rubbers or, to use the more modern term, elastomers derived from elastic polymers.

A number of the thermoplastics consisting of long chain molecules of great regularity are capable of aligning those chains very closely together. In this state strong forces of attraction can develop between any given chain and its neighbours. As a consequence individual molecules are restrained against movement under stress giving rise to hard, rigid material.

Other polymers possess chains whose shape and characteristics do not allow such close packing so that a more tangled structure develops in which limited movement of the chains is possible. In

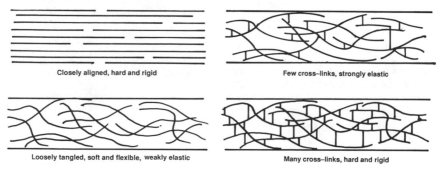

Closely aligned, hard and rigid

Few cross–links, strongly elastic

Loosely tangled, soft and flexible, weakly elastic

Many cross–links, hard and rigid

Arrangements of long chain polymer molecules.

this case a softer more flexible material results in which a degree of stretching under light loads is permitted.

It is the property of stretching under load then returning to the original length when the load is released that is termed elasticity. Some materials possess it to a very slight extent and others show very strong elasticity.

In the case above, the 'tangled' molecules are held in their positions relative to each other by fairly weak forces so they are only elastic under small loads. When larger loads are applied they tend to slip past one another to become permanently stretched instead of returning to the original position.

For a polymer to be strongly elastic it must not only have a somewhat tangled molecular structure, but those molecules must also be joined together at the points where they most closely fit in order to prevent any sliding under load. This is called cross-linking whereby chemical links are formed across from one chain to another at selected points. In rubber it is known as vulcanisation and the cross-linking chemical is sulphur. If too many cross-links are formed the material loses its elasticity and becomes hard and brittle. When too much sulphur is added to natural rubber the hard plastic *ebonite* is formed, which was made in large quantities before the discovery of modern polymers.

The mechanical properties of thermoplastics are very dependent upon the temperature at which they are used. This is of importance in understanding the behaviour of those adhesives like the anaerobics, which are of this type. At very low temperatures all thermoplastics are hard and brittle. As they warm up they become more flexible and further heating causes a change to a rubbery then a plastic state in which application of pressure alters the shape, as with damp clay. Finally, at a high enough temperature the polymer melts to a viscous liquid. The exact temperature at which these changes occur is different for each polymer, so it must not be assumed that their appearance at room temperature is the only state in which they can exist. The strength varies very widely as temperatures change.

Currently-used elastomers include natural rubber, the identical synthetic cis-polyisoprene, butyl rubber, nitrile rubber, neoprene rubber, ethylene-propylene co-polymer rubber, fluoro rubber, urethane rubber and silicone rubber.

Nitrile is a nitrogen containing co-polymer based on acrylonitrile, neoprene contains chlorine while polyurethanes contain both nitrogen and oxygen in the molecule. The silicone type is the most difficult to bond, although the forms in which it is used rarely call for such treatment.

METALS

Of greater interest to the model engineer is the making of strong bonds between various metal surfaces, between a metal and some other material, or the sealing of joints made with metal components.

The most commonly used metals as far as we are concerned are mild steel, and the basically similar carbon steels typified by silver steel, stainless steels which are alloys of iron with nickel and chromium, copper and its alloys with zinc or tin making up the brasses, bronzes and gunmetals, further alloys of copper with aluminium, often called aluminium bronzes, aluminium itself together with its alloys with silicon, magnesium, zinc, etc, and zinc alloys mostly encountered in the shape of die-castings.

Cast iron is a rather special case since it is manufactured in a variety of grades which differ mainly in the amount and form of the free carbon or graphite present in the finished casting.

It is generally accepted that the strongest bonds are made to clean, grit-blasted mild steel, with low alloy or carbon steels being almost as good. Good grades of cast iron are similar to the latter but need more care in the preparation. Stainless steels are high alloy materials in which the alloying elements modify the surface properties so that lower bond strengths result. Bonds to copper and its alloys are also weaker than those to mild steel. Aluminium and its alloys vary quite widely depending on the composition and type of surface treatment; anodised surfaces should be removed prior to bonding. Electroplated surfaces in general tend to give lower bond strengths, especially when very hard and smooth as with chrome, or if oxidised as with zinc and cadmium.

CHAPTER 6

JOINT PREPARATION

Disappointing results may occur in spite of careful selection of the correct adhesive for the material in question and proper joint design if attention is not given to preparation of the joint surfaces prior to application of the glue

We have seen that porous surfaces may be divided into those that are soft and flexible like paper, fabrics, leather, wood, etc or those that are hard and rigid like some ceramics, concrete, stone, and so on. Because they are porous they tend to absorb not only any liquids they come into contact with but also fats or greases as well as picking up dust and dirt particles. It is necessary then to make sure that all foreign matter is removed from the joint faces as a first step.

A particular problem arises with

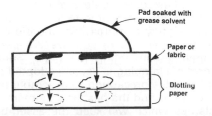

Removing grease stains.

paper or card in that attempts to clean it with water can lead to the material falling apart. If it is just dirty a wipe with a damp cloth might suffice; in more stubborn cases a little detergent could be added to the water, but wipe with clean water to remove traces of detergent afterwards. Grease or oil is best taken out by placing paper, or fabrics, on clean blotting paper then applying a grease solvent such as carbon tetrachloride, trichlorethylene or white spirit to the surface. The dissolved grease will pass through to be absorbed by the underlying paper. Repeating several times with a fresh sheet of blotting paper underneath and clean solvent each time is the correct method. Fabrics can be given a wash in clean soapy, or slightly detergent, water followed by a rinse in fresh water. Make certain the material is completely dry before placing the adhesive, as the penetration and subsequent setting may be adversely affected by a waterlogged substrate.

Leather should be treated with care for several reasons. The surface may have been finished by a thin coating of a flexible varnish-like substance which might be stained or damaged by some solvent adhesives; a trial with a small amount of the adhesive on a part not likely to be seen is the best insurance against this sort of problem. Secondly, the bulk of the leather may have been treated with an oil or similar substance to ensure it remains flexible, in this case too it might be difficult to get the glue to adhere. The remedy is to wipe just the joint surface itself with a cloth dampened with a solvent − do not swamp the surface with solvent which can remove all the dressing leading to possible brittleness. Leather, being essentially a protein product, is capable of being joined by a variety of adhesives, nevertheless care should be taken not to choose one that will make a rigid joint; a good degree of flexibility to match that of the leather should be the aim.

With wood the same rules of cleanliness apply. Contamination of the surface as found must be removed as well as old glue from previous joints that have failed. Old animal glue can be softened in warm water and the bulk scraped off, and a further short soak in warm water followed by a wipe with a clean cloth should extract most of the glue that had penetrated the wood.

When new joints are cut, carefully clean the tools of any protective grease that may have been applied when they were last put away, then make a few cuts through a piece of scrap wood. Adhesives work best in a closely fitting joint without gaps but if your skill does not run to making such a fit straight from the tool, complete one half of the joint to an acceptable shape and surface finish first. Make the other half slightly oversize for final fitting by trial and error by shaving down with chisels or even filing with a clean medium cut engineer's file kept for the purpose. Carpenters' rasps are too coarse for this

sort of work, being more suited to rough shaping before finishing by other methods.

Metal workers often fight shy of traditional woodworking preferring to machine the wood as though it were metal. This is fine as long as the machines, together with workholding devices and cutting tools, are thoroughly cleaned before starting. Clean paper or card between work and holding surfaces is a great help. End mills or slot drills enable accurate mortices to be cut, the rounded ends are no problem since the mating tenon can be quickly rounded off to match by filing. When cutting across the grain use the woodworker's trick of marking the line by a sharp knife cut to avoid ugly splintering.

There are many instances where it is necessary to stick pieces of wood together, or to other materials; even the metal worker sometimes needs to make a box, a showcase, a base for a model or perhaps a pattern from which to produce a metal casting. Practice is the best way to learn to make the most suitable type of joint for each job, to provide the degree of accuracy for a sound adhesive joint and to appreciate which adhesive best suits the task.

Having made a satisfactory joint there is one more step to take before rushing to apply the adhesive. Remember that wood is porous so it will take time for any solvent to evaporate; even with solventless glues, unless of the instant type, it also takes time to set. Wood is also slightly compressible, therefore it is an advantage to squeeze the parts together eliminating the gaps. Some form of clamping is usually adopted so it is a wise precaution to assemble the job dry to find the best arrangement of clamps which will keep the assembly 'square' during setting. Sometimes one

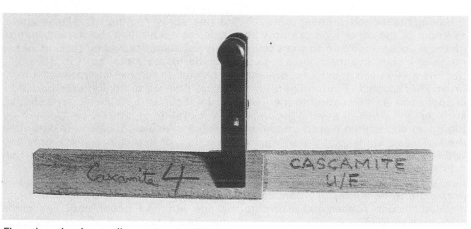

Firm clamping is usually necessary with most adhesives.

has to make wedges or shaped clamping blocks in order to prevent slipping — it is too late to find this out after the glue has begun to set.

Rigid porous materials can be washed or treated with solvents to shift contamination, but must be given adequate time to dry out afterwards. This can be hastened by warming or by the blast from a hair drier or hot blower. Because hard brittle materials break with an uneven surface, a trial fit should be made to ensure a close joint can be obtained without gaps. If pieces have broken away it is essential to find where they fit and whether they can be fitted in after the main pieces have been assembled. If not then a definite order of assembly must be decided upon.

Some porous materials, both rigid and flexible, have such an open structure that all the adhesive might be absorbed into the interior leaving little at the joint surfaces. There are two ways to overcome this, the first involves the application to each surface of a thin coating of adhesive which is left to air dry until it is almost set, when a second coat is given to one of the surfaces before closing the joint. Sometimes the first coat is thinned somewhat if water- or solvent-based. Technically this method should be referred to as a sealing operation and the first coat as a sealing coat, but it has been the practice to call it a priming coat. Unfortunately another type of product is also called a primer, which is of a different composition to the adhesive with which it is used but improves the adhesion of it. The second method of dealing with a very porous material is to use an adhesive that has thixotropic properties. This means that it has been formulated to be used as a thick gel which will not run or spread, in other words it stays put on the joint surface instead of soaking right in.

NON-POROUS MATERIALS

Plastics, except for some open cell foams, may be regarded as non-porous materials for the purpose of making adhesive joints. When in fibre form, they give a porous substrate which should be treated like other fabrics.

Having mentioned 'priming coats', an example of the other type can now be given. A recent addition to the Loctite range of instant cyanoacrylates is Loctite Prism 406 used in conjunction with Prism 757 Polyolefin Primer. The primer is applied as a thin coating to the joint surfaces and dries in several seconds, the Prism 406 is then placed on one of the surfaces and the joint closed, again, within seconds, bonding has occurred.

Prism 757 Polyolefin Primer enables the 'polyolefins' polythene and polypropylene to be succesfully bonded as well as silicone rubber and claims to also bond to PTFE.

The surface preparation of plastics needs some care due to the manufacturing processes employed, some of which can leave a degree of surface contamination although the material might look clean.

Injection-moulded components are formed by forcing the hot molten plastic into metal dies at high pressure. To facilitate withdrawal of the moulding from the die when cooled the metal has

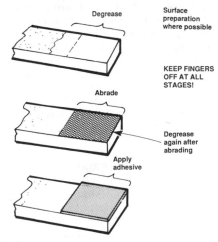

Surface preparation.

68

a fine spray coating of 'release agent' prior to each filling. By its very nature this substance prevents 'sticking' of the plastic to the metal so it is obviously essential to remove all traces of it from areas that are to be adhesive bonded.

Joint surfaces on plastic parts should, as a matter of course, be cleaned by wiping with a cloth dipped in detergent solution, or in trichloroethane provided the material is not attacked by that solvent.

This treatment should be followed by mechanical abrasion using a medium grade abrasive paper until any shine has been taken off. Compared to a smooth surface an abraded one has a greatly increased bonding area by virtue of all the vee-shaped scratches covering it. In addition a freshly abraded surface has increased affinity for the adhesive molecules, hot-moulded surfaces tend to be rather inactive.

Adhesive should be applied as soon as possible following abrasion, being careful to avoid touching the joint surfaces with the fingers during handling as grease is frequently transferred in this way.

Another possible contaminant that might exist on new sheet plastics is traces of adhesives left behind when the protective packaging layer is stripped off, the same pre-treatment as previously mentioned should ensure its removal.

Migration from some plastics of the plasticisers added to render them flexible can cause failure of adhesive bonds – PVC often contains them. A wipe over the bond area with a clean cloth dipped in solvent helps to remove this source of trouble.

Absolute cleanliness of tools and equipment liable to come into contact with the material is a must when dealing with plastics, as it is with other substances being prepared for gluing. Make a habit of covering the bench with clean paper whenever this sort of work is undertaken and the failures should be far fewer.

METALS

It is usually difficult to avoid contamination by oils or cutting fluids when working with metals, though brass does seem to be easier to keep clean than steel or aluminium for instance. Fortunately metals are more readily cleaned than most other materials so perhaps it does not matter too much.

Since metals are little affected by the temperatures at which solvents boil, and since most solvents do not attack them, industry makes much use of chemical degreasing baths. In these the metal object Is suspended above a tank of boiling solvent which causes the vapour to condense on the cooler metal, where it dissolves any grease or oil and runs back into the tank. Fresh vapour free from oil is constantly condensing so that in a very short time the metal is clean. Excess solvent is prevented from escaping by a cooling coil placed around the top of the tank.

Although non-inflammable solvents like trichloroethylene are used to eliminate fire risks there is a toxicity hazard to workers constantly exposed to the small amounts of vapour released when parts are removed from the tank, and the system is not easily adapted to small scale occasional operation in the home workshop.

It is best to get rid of large amounts of contaminant by washing the metal with ordinary paraffin, wiping away the paraffin with a cloth then giving a wipe over with another clean cloth dipped in 'trike' or another volatile solvent.

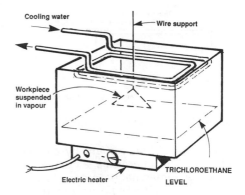

Vapour degreasing tank.

Always ensure there is adequate ventilation when using any form of solvent indoors, making sure all flames are extinguished too, even when the solvent is non-flam, because many solvents are decomposed by heat to produce poisonous fumes.

Immediately before fitting the parts together give yet another wipe with a solvent-dampened tissue to the mating surfaces and abrade lightly with medium grit paper.

When attempting to stick parts to unmachined surfaces such as rolled sections or castings all scale, rust and paint or other protective coating must be removed as a first step. The joint area should then be filed, chiselled or scraped flat for the parts to fit closely together. Unless this is achieved a lower strength joint will result. Where some pitting or unevenness is unavoidable it is necessary to use a special gap-filling grade of compatible adhesive, but in general thick glue lines give lower strength than thin ones.

Although degreased and abraded or grit-blasted surfaces provide a good basis for adhesion it is possible to ob-

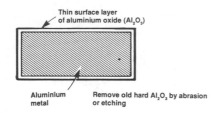

Thin surface layer of aluminium oxide (Al₂O₃)

Aluminium metal

Remove old hard Al₂O₃ by abrasion or etching

Anodised aluminium.

tain further improvement by chemical etching. The chemicals employed are however somewhat hazardous to use in the home environment as they usually contain one or more of the following acids: chromic acid, hydrochloric acid, nitric acid, phosphoric acid or oxalic acid, depending on which particular metal it is intended for. This is really only an industrial process where the very highest bond strengths are essential, for example anodised aluminium often has a poor surface for adhesives to attach to and is etched with chromic acid solution or sulphuric acid/dichromate/chromium trioxide solution which amounts to the same thing; the effect is to strip off the anodised oxide layer leaving an activated metal surface. Almost as good results can be gained by grit-blasting which also removes the anodising and with care scraping or use of abrasive paper will produce a satisfactory finish.

Cast iron in its machined state can be difficult to stick because of the graphite that is exposed during cutting and which is a well known lubricating agent that gives the material its excellent wear properties. The graphite powder can be slowly released over a long period. A lubricant is the last thing we want in an adhesive joint so good preparation of the surfaces must be a priority. A very good way is to scrub the surface with a stiff, fine bristled brush in hot detergent solution, changing the water several times until it no longer becomes black. Then rinse well in clean hot water with no detergent and dry quickly before the metal has a chance to rust, which can happen in a surprisingly short time under these conditions. To ensure that any very slight film of corrosion does not impede adhesion, rub the adhesive vigorously onto the surface with a piece of clean hard wood to break up and disperse the layer.

Glass is one of the more difficult materials to bond with adhesives partly on account of its chemical composition and partly because of the very smooth surface. There is fortunately one easy way of deciding when a glass surface is free from contamination. As you will have noticed, a car windscreen that has been dry for some time becomes smeary when it first begins to rain because the traffic fumes have made it greasy. Once the wipers have been in operation for a while the glass clears, since removal of the grease allows the water to 'wet' the glass forming a continuous thin film instead of standing in separate droplets. Alcohol or detergent in the windscreen wash bottle speeds dispersal of the contamination.

It is this property of water being able to wet the glass that enables us to decide when it is clean. Therefore wash any glass article to be bonded either with a solvent or with warm water containing a detergent, rinse off in fresh water and notice whether the whole surface remains wet. If the water layer breaks up into streaks or drops the glass is not clean and the washing must be repeated.

When clean, abrade the actual area of the joint faces with medium grit paper or by masking and sandblasting, wash all dust away then dry by warming or

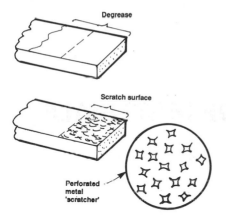

Degrease

Scratch surface

Perforated
metal
'scratcher'

Rubber surfaces.

blowing with hot air. Since water adheres very strongly to clean glass it is best to heat the article to some ten or twenty degrees above the boiling point of water. Be careful to heat the glass slowly and evenly to minimise expansion stresses that might cause cracking.

Apply the adhesive to the joint area while it is still warm but only in a thin layer in order to prevent runs.

Rubbers or elastomers, like plastics, need treating with solvent to get rid of anti-stick or mould release agents, but do not soak the material as some solvents cause softening and swelling. An alternative is to wash with detergent and water which is probably the safer method.

Abrading rubber is not easy with normal abrasive papers, but the perforated metal scorers supplied with rubber 'stick-on' soles and heels are quite effective. The aim should be to obtain an even, complete coverage of the bond area paying particular attention to the edges where 'peel' is a problem.

If joining rubber 'O' rings or belts it is not really feasible to abrade such small surfaces, but a clean cut with a sharp knife at right angles to the length will enable a joint to be made with one of the engineering grade instant adhesives designed for the purpose.

CHAPTER 7

IDENTIFICATION OF MATERIALS

Before being in a position to decide upon a suitable adhesive for any particular job, you need to be certain of, or at least have a good idea of, what material it is that has to be bonded. In some cases this will be fairly obvious, glass for instance is reasonably easy to distinguish from the various transparent plastics. It is not so easy however to decide which polymer of the several transparent plastics it may actually be. Indeed it is in the group of plastics as a whole that the greatest confusion is likely to arise since different polymers may have the same colour and physical appearance but have totally different bonding behaviour.

Experience in using the materials goes some way towards aiding identification, but without carrying out certain simple tests considerable doubt must still remain. These tests include comparison with known samples of different polymers, flexibility or hardness, transparency, effect of heat, colour of flame when a small piece is burned, whether it continues to burn when removed from the flame, if it drips and whether the drips continue to burn, the type of smoke if any, and the smell produced by burning.

Taking the last point first a word of caution is necessary here since the fumes from hot or burning plastics can be quite toxic. Bear in mind that we are talking about burning *small* pieces not larger than half a matchstick, and it is only one end of the piece that is actually touched to the edge of a *small* non-smoky gas flame. When testing for smell it is not necessary to completely inhale the fumes; the cells that detect odours are in the upper part of the nose so it is only required to *gently* sniff a *little* of the fumes and almost immediately blow them back out. To deeply inhale a large amount of fumes is not only dangerous but it overpowers

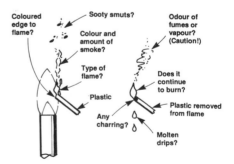

Identification of polymers.

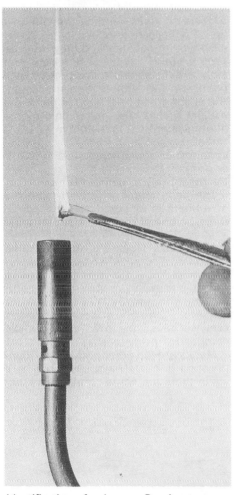

Identification of polymers. Burning test on PVC.

collection of known samples of different polymers and to check that one of them burns in the same way as the unknown. Then, even if it is a different colour, you can be pretty sure it is the same or a very similar polymer.

Let us assume we are identifying a transparent plastic and look at the way it might react to both gentle heating and to burning. A piece of the old fashioned celluloid (some of which may still be lying around, as the set-square used to make the accompanying drawings is of this material), if warmed, does not melt and drip; in fact it very easily ignites and burns extremely rapidly with a non-smoky, slightly yellow flame. No modern plastic burns like that, it would be considered too dangerous.

Turning now to its replacement, which is still made, we can look at the behaviour of *cellulose acetate*. This is readily available in the form of photographic film negatives which consist of a thin sheet of the plastic supporting a very thin layer of gelatin containing the silver image salts. Soak a piece of exposed negative in warm water until the gelatin softens and can be removed, wash well to get rid of all traces of it.

When heating at the edge of the flame hold the end of the strip in tongs or tweezers. It does not melt but begins to char and only burns if quite strongly heated, then it will continue to burn with a smoky flame though not very vigorously. The smell of the fumes before it begins to burn is faintly like vinegar. Although known as safety film it is not completely non-flam but is far safer than the cellulose nitrate that was previously used.

Perspex, or *polymethylmethacrylate*, softens when heated and burns vigorously with a clear light yellow flame without smoke. When the flame is

the sense organs leading to reduced sensitivity. Needless to say, remove all solvents and inflammable material from the vicinity before starting these tests.

It is a very good idea to make a

blown out the vapour has a pleasant fruity smell. Other acrylics are somewhat similar but the odour is different.

Polycarbonate burns if heated strongly, it has a smoky flame and chars heavily. There is little smell.

Polystyrene softens and burns with an extremely smoky flame producing black smuts that float around in the air. The fumes when it is blown out smell rather sweet and sickly.

Polythene melts and burns with a non-smoky flame, drips of melted material fall from the flame. The vapour smells waxy.

It is not easy to describe odours to other people which is why it is recommended that you carry out your own tests but do take the precautions previously outlined and do not attempt to hold the pieces in the fingers, use tongs or tweezers every time.

These tests can of course be made on other plastics that are not transparent, here are details of a few more:

PVC, or *polyvinylchloride*, shows a green edge to the flame and burns with a smoky flame with charring. It has a typical choking smell.

Polyurethane softens, chars heavily and burns while in the flame with a slight coloured edge to the flame. Its fumes have an acrid smell.

Nylon melts, bubbles and chars slightly before it burns with molten drops falling. The smell is like burning hair or wool.

PTFE softens and shows a green edge to the flame but does not burn. *Do not* sniff the vapour when you see this happen as it is very toxic.

This method is probably the best way for the amateur, or indeed for anyone not having access to sophisticated techniques of chemical analysis, to attempt the identification of polymers including rubbers. It does require the building up of a collection of small pieces of each type of polymer and they must be kept properly labelled. Stockists and suppliers listed in *Yellow Pages* will often help with offcuts for a small sum.

A slight green coloration to the flame usually means that either chlorine or fluorine is present in the polymer, but it can also result from traces of copper in certain dyes or pigments used to colour or print on the plastic; blue colours are the most likely to contain copper.

The best source of known samples of rubbers is to purchase labelled 'O' rings, as one of these will last for quite a long time if carefully used in the burning tests. Natural rubber has quite a distinctive odour when burned; everyone must know the smell of burning rubber, it is easily possible to distinguish between this and nitrile or neoprene for instance. Fluoro and silicone elastomers likewise have their own characteristic combustion properties.

Metals are the other area where identification is not always easy, but here again there are some simple tests that

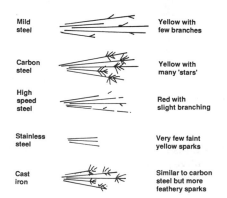

Identification of ferrous metals.

74

can help to decide the category if not the actual metal or alloy itself.

Colour of the metal is a fairly good guide which will at least indicate to what group the unknown sample belongs. Of course any dirt, corrosion or surface coatings must first be cleaned off to give a view of the underlying metal. Copper is a well known reddish colour as are some of the high copper alloys including some gunmetals. The brasses usually have a distinctly yellow tint which approaches a golden appearance in the high copper variety called gilding metal. The 60% copper/40% zinc type of brass known as yellow metal has a deeper ochre colour and is also called muntz metal.

Aluminium and its alloys have their own distinctive colour which is not easily mistaken for other 'white' looking metals, in addition its softness and light weight confirm the identification.

The steels are not always easy to differentiate without a few simple tests, the first of which is to see if the sample is attracted by a magnet. The only common type of steel that is not magnetic is the 18% chromium/8% nickel grade of stainless steel, also called austenitic stainless. The other types of stainless steel are magnetic – they are the sort containing about 12% chromium without nickel called martensitic stainless and the chromium iron or ferritic stainless. These latter two do not have the same degree of corrosion resistance as the austenitic variety but are still much better than mild steel. To distinguish them from mild steel just degrease and leave the sample on a piece of damp newspaper overnight – mild steel will show unmistakable rusting.

An ancient method of identifying ferrous metals is the 'spark test'. The sample is held with its edge pressed lightly against a rotating emery wheel so that grinding sparks are thrown off. The colour, thickness, length and degree of branching of the spark are quite characteristic for each type of steel.

This is another case where it is wise to keep a collection of small labelled pieces of the various steels in order that the grinding sparks may be readily compared. The following is an indication of the differences:

Mild Steel light yellow, long straight sparks with a few small branches.

Silver Steel yellow, many-branched sparks.

High Speed Steel red unbranched sparks.

Stainless Steel very few thin yellow sparks.

Zinc die-cast alloys may look superficially like aluminium alloy when new and polished but soon lose their shine to become a somewhat dirty grey colour. By comparing with a known sample of each it is usually not too difficult to decide to which type the unknown belongs.

It was said earlier that surface deposits should be removed so that the colour of the metal can be seen, but there is one form of deposit that often gives an idea of what lies underneath anyway. Corrosion products are indicative of at least the predominating metal in many cases, for instance a brown or red rusty appearance is a good guide to the ferrous metals except for stainless grades. A green deposit indicates that copper is present, though nickel plating can show a light green but it should be possible to recognise this with a little practice. Aluminium and zinc alloys both give white deposits.

While on the subject of making collections of materials as an aid to identification we should not forget timber varieties. Small squares or rectangles of planed wood all made to the same size and thickness can be glued onto a piece of hardboard and framed just like a picture. With neatly typed labels it would not look at all out of place hanging on the workshop wall.

Electroplated metals pose bonding problems either because of the poor affinity of the metal to the adhesive, as with zinc- and cadmium-plated articles, or by reason of the very smooth surface finish that is typically seen with chromium or nickel plating. The ability to identify the presence of such coatings will then affect the efficiency of the bonding.

Cadmium plating was very popular at one time as a corrosion resistant treatment for small steel articles. It is often of a fairly matt, slightly tinted with a mixture of colours, irridescent appearance. To a large extent it has been phased out in recent years in view of its toxicity, either through being handled and picked up on the skin or by the fumes being inhaled when it is heated. If present it is best removed from the bond surfaces by abrasion with medium grade abrasive paper, used wet to avoid the risk of picking up or inhaling the metal dust.

Zinc has now replaced cadmium plating in most areas where it was formerly common. Zinc plating when new has a shinier appearance with a whiter look than chrome plate though not as glossy as the latter. When weathered it becomes dull looking and may develop whitish deposits. Again it is best removed by abrasion – use sellotape to mask off the area around the joint faces while abrading if the appearance or corrosion resisting properties of the finished part demand it.

The much harder chromium plating is so well known as to be easily recognised by all; it is usually very smooth with a slight bluish tint. Being difficult to remove, it is best abraded using silicon carbide paper which has a more wear-resistant grit than the normal types. Aim to achieve a matt finish with the high gloss just taken off.

Nickel plating has a 'tinny' look to it with a colour more like that of silver. It too is quite hard and should be lightly matted.

Tin plating itself as seen on 'tin cans' is soft enough to be readily abraded away to reveal the underlying metal.

Painted or varnished surfaces are not difficult to identify. They represent a poor surface for adhesives because the bond between substrate and paint is very weak compared with the adhesive bond we are trying to make. Scraping or abrading away the paint layer, with masking to avoid damage to adjacent areas, is considered essential.

In cases where a previously made bond has failed, examine the bond surfaces to determine the mode of failure. It will soon be noticed whether the adhesive has peeled from one or other of the surfaces showing poor affinity or improper preparation. Paint or other material that has pulled away with the adhesive will also be apparent. This is another way of identifying unsuitable surfaces.

CHAPTER 8

ENVIRONMENTAL FACTORS

Since adhesives are organic compounds we would expect them to be affected by the normal environmental conditions of temperature, humidity, ultraviolet radiation, oxidation by the oxygen of the atmosphere and, in some cases, bacterial or fungal attack. Most people seem aware nowadays of the presence of ozone in the atmosphere. This very reactive form of oxygen is produced at high altitude by the intense radiation from the sun, but the layer of ozone absorbs much of the UV radiation that would otherwise reach the ground and so has a beneficial effect. Some ozone is formed at ground level too, especially in 'smog' conditions whereby sunlight causes complex chemical reactions with pollutants in the atmosphere. This low level ozone accelerates the deterioration of organic matter and is particularly detrimental to some rubbers. When a stretched piece of rubber is exposed to ozone it rapidly develops a mass of cracks at right angles to the stress. This phenomenon is known as 'ozone cracking' and a variety of chemicals is added to rubbers to reduce the risks. Waxes are fairly effective but, as might be

Example of ozone cracking on rubber drive belt.

77

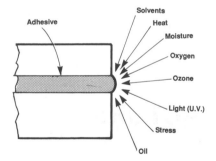

Environmental factors that affect adhesives.

guessed, this practice does not help adhesive bonding.

In fact a whole range of chemicals is added to polymers in an effort to slow down environmental degradation. It cannot be completely prevented and in time virtually all organic materials are broken down otherwise all the waste would accumulate and life would grind to a halt. Polymers, then, consist not of pure substances but are a witches' brew of anti-oxidants, ozone inhibitors, UV absorbers and so on, together with substances to ease the mixing and forming or to modify the physical properties. Cookery recipes are quite simple compared to the ingredients that go into some polymers, particularly rubbers.

Incidentally, appreciable quantities of ozone are produced by electrical discharges especially at high voltages or in the presence of sparking. This is said to account for the peculiar odour in the vicinity of working dynamos etc. Rubber used on or near such machinery is susceptible to attack so it is specially compounded to minimise the risk.

In the field of wood adhesives we have already seen that the traditional animal glue is capable of making strong joints; if correctly made and applied, the bond is actually stronger than the

wood itself when kept dry and at room temperature. The effect of moisture is to cause a considerable reduction in strength or even complete failure. Moisture also allows bacterial and fungal damage to occur which destroys the glue in time as well as weakening the wood.

Some of the synthetic adhesives, particularly those based on cellulose or other natural substances, are also susceptible to softening by water, some contain fungicides to prevent weakening or discolouration as well.

There are few truly waterproof adhesives for wood. Ones that will withstand constant exposure to the atmosphere or constant immersion in water under extreme conditions are labelled BWP which means Boil and Water Proof. These adhesives are tested by using them to glue pieces of wood together and then boiling the assembly in water. This is a very severe test indeed but is obviously of importance in providing an assessment of the likely durability of the product. It should be noted that it does not mean the glue is intended to be used for articles submerged in boiling water, it is really an accelerated aging test that indicates the product has the capability to withstand severe conditions.

In the realm of boat building or repair such a glue would obviously be required, but there are few candidates that fit this specification. Undoubtedly one of the best and relatively easy to use is resorcinol/formaldehyde which is a two-part adhesive consisting of liquid resin and powder hardener. Mixed just before using, it is a dark red colour that shows up the glue line.

Another good adhesive is phenol/formaldehyde but this is not so easy for amateur use. It is widely employed by

Urea/formaldehyde adhesive (left), resorcinol/formaldehyde resin and hardener (centre), and PVA co-polymer emulsion for outdoor use (right).

industry including the manufacture of 'marine plywood', a material that must be capable of withstanding the BWP test.

Epoxide adhesives also have good waterproof properties as well as adequate strength when used to bond timber.

Usually if no reference is made on an adhesive label to the effects of humidity or moisture it should be assumed that it would be affected by these conditions. Adhesives that may be used in occasional humid or damp conditions, but not exposed for a prolonged period, are properly labelled water resistant. Waterproof implies that wet conditions can be withstood for an appreciable time but does not mean that continuous immersion or prolonged dampness will not cause the bond to deteriorate.

Although not of the BWP grade, urea/ formaldehyde adhesive is fairly close to it and has been used in the building of boats from marine plywood. It is easy to mix and use giving a white glue line.

Where an adhesive like a rubber-based contact variety is used to fasten a laminated sheet to a worktop, the impervious nature of the surface will ensure that moisture is unlikely to reach the glue, and will also protect the glue from ultraviolet light, oxygen and ozone, etc. This will be true of many other situations where adhesives are used so that any attack will be limited to the exposed edge of the glue line. Moisture penetration is however very insidious and will exploit any region of weakness; its ability to creep along the bond surfaces should not be underestimated.

The effect of heat is to cause a softening of organic substances at moderate temperatures, usually in the range of room temperature up to about 200 degrees Celsius. This softening leads to a loss of bond strength, the amount of which depends upon the chemical nature of the adhesive. Above this temperature some degradation of the adhesive will tend to occur, with permanent damage and eventual destruction of the bond. The rate of chemical reactions doubles for every ten degrees rise so it is inevitable that environmental attack becomes more severe in heated conditions. Manufacturers often state both minimum and maximum temperatures between which the particular product may safely be used.

The effect of low temperatures is to make substances become more and more rigid, again this varies with the types of adhesive but there is a point at which the behaviour changes from elastic to brittle leading to greatly reduced bond strength. Even rubbers will lose their elasticity if cooled to a sufficiently low temperature; neoprene, for instance, being usable down to about minus 40 degrees Celsius. This change of properties is sometimes referred to as the 'glass transition' temperature below which the material will fracture in a brittle fashion rather like glass, although the fracture at higher temperatures may be quite ductile.

Similarly the change from elastic properties to plastic behaviour can be called the plastic transition, thus a material like nylon for instance will be tough and springy at room temperature, and for some way above this point, but in the vicinity of the melting point at about 260 degrees Celsius it is quite plastic so will suffer a permanent change of shape when stressed.

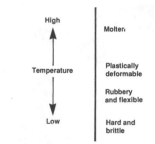

Effect of temperature on thermoplastics.

Thermoplastics are more sensitive to temperature changes than thermoset materials, nevertheless adhesives of the two-part type still have limits outside which their bond strength falls off significantly.

Other substances with which adhesives may come into contact include a variety of chemicals, the most common of which might be oils. The type of glues most likely to suffer in this case would be those having an affinity for that sort of oil, whether it be vegetable or mineral in origin. Motor oil for instance rapidly softens certain rubbers, including natural rubber, so adhesives based on them would be weakened.

Solvents too may be present in liquid or vapour form in some areas where an adhesive joint is required. This possibility must be considered when choosing a product and guidance is often given on the packaging or data sheet. Where more detailed advice is needed manufacturers will often make recommendations on receipt of a request accompanied by a description of the substrates and the environment to which the joint will be exposed; joint configuration and area, together with desired strength or estimated stresses, should be included.

CHAPTER 9

ENGINEERING ADHESIVES

It is only in recent times that adhesives have been applied to engineering problems on a large scale. Today there is a wide range of products to cover a multitude of different tasks ranging from threadlocking to the bonding together of complex structures. Without an understanding of the terminology it is difficult to choose a product suited to a particular job – to choose the wrong one often leads to poor results and a lack of faith in the method. As a first step some definitions of the terms will be given. Those in general use are: threadlocking, pipe sealing, liquid gasket, retaining compounds, instant adhesive, structural adhesive.

THREADLOCKING COMPOUNDS

These substances are not true adhesives and have little adhesion to the surface of the threads. They consist of anaerobic liquids that polymerise in the absence of air, a process that may be accelerated by the extremely thin film of oxide always present on the metal. The solid polymer that forms in the gap or clearance space between the mating threads has the effect of increasing the friction, so requiring more force to unscrew the components. In this respect they act in a similar manner to the nylon ring trapped in the top of a 'Nylock' nut which reduces the risk of loosening under vibration. Threadlocking compounds are applied as a single drop of liquid to one of the threaded parts just before assembly and set within a few minutes of tightening the screw, nut or stud. They replace mechanical spring washers and other locking devices.

A number of grades of threadlocking compound are produced so that a range of strengths is available to suit different uses. For instance, stud locking grade is stronger than nut locking grade so that a stud secured in this way will not unscrew when the nut is removed, the stud itself can be unscrewed either by the usual method employing two nuts jammed together or by gripping the plain shank if one is provided.

Since the locking compound has the effect of spreading the load over more of the threads, stress concentrations are reduced giving improved fatigue life. Plain studding, or 'all-thread', may be safely used instead of specially machined expensive conventional studs.

An added advantage of locking com-

81

Numerous studs, nuts and screws secured against vibration with anaerobic sealing compound.

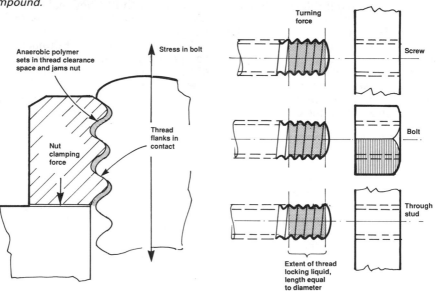

Anaerobic polymer sets in thread clearance space and jams nut

Stress in bolt

Thread flanks in contact

Nut clamping force

Turning force

Screw

Bolt

Through stud

Extent of thread locking liquid, length equal to diameter

Anaerobic threadlocking action.   *Application of threadlock.*

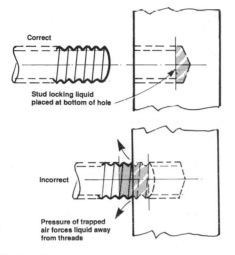

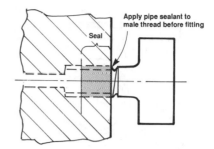

Sealing and locking fittings.

Fixing blind studs.

pounds is that the thread is effectively sealed against the entry of moisture thus reducing the risk of corrosion which often causes difficulty when dismantling. Thread breakages and the problems of removing broken studs are largely eliminated.

Where bolts are used in inaccessible places one of the stronger threadlocking compounds or even a retaining compound may be used to keep it in place and prevent rotation while the nut is tightened.

PIPE SEALING COMPOUNDS

These are similar in many ways to the threadlocking compounds but have been formulated to withstand pressure that might otherwise force the sealant from between the threads. Not only do they form a pressure tight seal, but also lock the fittings in position preventing loosening due to vibration. Again, they protect the threads from corrosion both

by the fluids within the pipework and externally by environmental conditions.

Dismantled joints are readily cleaned of old sealant by wire brushing or running a die-nut down the thread ready for rebuilding with a fresh application. They seal against liquids, gases and vapours but should not be used in contact with oxygen supplies as all oils, greases and other organic substances are banned in this particular case.

LIQUID GASKET COMPOUNDS

There are two classes of product in this group. The first comprises acrylic ester anaerobic variations, and the second type is the silicone rubber that sets in the presence of the moisture normally adhering to surfaces in an invisible film.

Anaerobic gasket compounds are supplied as a one-part liquid which is applied to one face of a flat joint area. Being thick liquids they can be squeezed from the container as a continuous bead around the joint area (it is not necessary to spread the compound all over the surface) so that when the joint is closed a thin film is formed which then sets. The joint is normally pulled up tight to give virtually metal-to-metal contact, any small gaps or scratches are filled up and sealed. It will

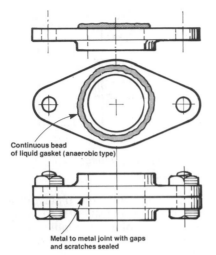

Continuous bead
of liquid gasket (anaerobic type)

Metal to metal joint with gaps
and scratches sealed

Application of liquid gasket.

be appreciated that with flat, well fitting surfaces only a small amount of compound is required. The products will however fill gaps up to about 0.015 inch or so and have a degree of flexibility to accommodate expansion- or vibration-induced small movements.

Liquid gaskets can be used in situations where conventional cut gaskets would be difficult to make or fit, for example with multiple faces, narrow flange faces or numerous bolt holes. In these cases application by roller or similar technique will ensure even and complete coverage without a large excess that might find its way into the system being sealed. Because these compounds have only low adhesive strength the joints are readily separated for servicing or repair, the old com-

Small locomotive boiler fittings sealed and positioned using anaerobic thread-sealing compound.

84

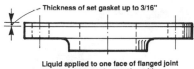

Thickness of set gasket up to 3/16"

Liquid applied to one face of flanged joint
and allowed to set by reaction with
atmospheric moisture before closing joint

Silicone sealant gasket.

pound is then easily removed ready for reassembly with a fresh coating.

The silicone rubber type is applied to one joint surface in an even film and reacts with moisture from the atmosphere to set into a solid flexible gasket. The thixotropic nature of the compound means that it stays put on the flange surface and may be built up to several millimetres thickness if desired. Since it is allowed to set before the joint is closed it can serve as a conventional gasket where metal-to-metal contact is not a requirement.

RETAINING COMPOUNDS

Where concentric components have to be fastened together permanently without the use of mechanical fastenings a retaining compound may be employed. Once more these substances do not rely on the formation of a strong adhesive

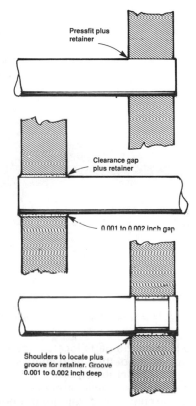

Pressfit plus
retainer

Clearance gap
plus retainer

0.001 to 0.002 inch gap

Shoulders to locate plus
groove for retainer. Groove
0.001 to 0.002 inch deep

Wheel and axle joints, etc.

Morse taper arbor having mounting boss attached with high strength anaerobic retainer.

Tender axle secured in wheel boss by high strength anaerobic retainer.

bond to keep the parts located, instead they make use of their high static shear strength to resist the stresses.

Where tensile shear strength refers to the strength of a lap joint under tension and comprises both adhesive and cohesive properties of the adhesive, in concentric assemblies it is the resistance to deformation of the solid polymer trapped in the gap between the surfaces that determines the strength.

There are two ways in which these compounds function: one is to resist forces in an axial direction tending to separate the components, while the other resists torsional forces or the tendency to rotate around the joint line.

Retaining compounds are anaerobic substances which are applied to the joint surface in a liquid state and polymerise when the joint is closed. A range

of strengths is available to suit different applications, the one chosen depending upon design constraints and the loads expected.

Typical uses for the high strength grades include the securing of wheels, pulleys, gears, sleeves, and other components, to shafts, the building up of shouldered shafts from pieces of different diameters to avoid wasteful machining, securing of liners of inserts, built up crankshafts, and so on. A weaker grade would be used to retain bearing bushes or races in housings, or for the fitting of parts that may have to be replaced at intervals.

Since these compounds do not entirely rely on adhesive strength they can be used to join parts made of different materials. Where dissimilar metals are joined the insulating film

Locomotive crankshaft built up from parts secured with high strength anaerobic retainer.

of polymer between the surfaces considerably reduces the risk of galvanic corrosion occurring as well as resisting environmental corrosion damage.

Although these compounds will fill gaps usually up to two or three thousandths of an inch, it is not an essential requirement that there should be a gap of that order. It does mean however that reamed holes and shafts turned to very tight tolerances may be dispensed with, but this does not mean that sloppy fits or rough machining should be tolerated as it can lead to problems of alignment or concentricity. There are three basic methods of fitting concentric parts using retaining compounds: the first consists of a *light* press fit in which the parts are assembled after wetting each component with retaining compound — the advantage

of this procedure is that the lubricating effect enables fitting under lower pressure and therefore with less risk of distortion or breakage than with

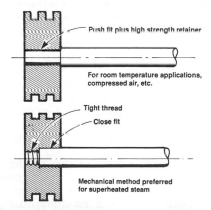

Piston rod fastenings.

Ball race secured in aluminium alloy housing using anaerobic retaining material.

really accurate results matter. To some extent this will depend upon the length to diameter ratio of the joint since a short one can tilt more than a long one. There is a self centring tendency due to capilliary action when one of the parts is very light, but this cannot be taken for granted with heavier pieces. Whether to support the assembly horizontally or vertically must be judged in the light of its dimensions and symmetry.

With all three of the above variations, it is permissible to add extra strength to the joint by means of keys or cross-pinning in those cases where limitations of space do not allow sufficient joint area to be certain of the retainer accepting maximum loadings. It is recommended in such instances that the parts should be fully machined before applying retainer, then the key or pin can be inserted before it sets so that a continuous film of polymer completely fills all gaps. If for any reason it

ordinary press fits. Accurate concentricity is assured, strength is enhanced and subsequent dismantling, if required, is easier because corrosion or seizure within the joint is prevented. The second method starts with a good push fit between the parts but then a shallow groove is turned in one or other of the components to a depth of only one or two thou over most of the joint area but leaving a short zone of contact at each end. The parts that fit locate the pieces accurately, leaving the groove to be filled with retainer as the assembly proceeds.

In the third method a gap of about two thou is left between the parts over the whole of the joint area. It is necessary to support the parts carefully during setting of the retaining compound if

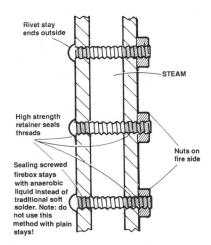

Rivet stay ends outside

STEAM

High strength retainer seals threads

Nuts on fire side

Sealing screwed firebox stays with anaerobic liquid instead of traditional soft solder. Note: do not use this method with plain stays!

Boiler stay sealing.

should be necessary to drill and pin after assembling the components, sufficient time must be given for the retainer to fully harden before commencing to drill – the pin can then be inserted after coating with retainer.

Designs that include properly calculated joint surfaces should be quite capable of accepting expected loads without any other form of fastening. Most of us, particularly amateurs, have a belt and braces attitude that is not always mistaken; it is often wise to be somewhat cautious, particularly where safety is concerned, so do not feel condemned on that count. It is precisely because it is not always possible to estimate the stresses to which an assembly might be subjected that an element of overdesign is desirable in addition to any factors of safety included in the calculations. The fact that the experts sometimes get it wrong and have embarrassing failures should put us on our guard.

As already mentioned locking, sealing, gasket and retaining compounds are not intended to be used as adhesives for tensile or tensile shear joints, since they do not have sufficient adhesive strength for these applications. The products that follow are true adhesives.

INSTANT ADHESIVES

This type is also known as 'cyanoacrylates' which are one-part liquid products that set in the presence of atmospheric moisture. The setting time depends on the particular substrate to which it is applied, as well as to the actual grade of adhesive that is used, but is always measured in seconds. The times can range from less than 5 seconds for a fast grade on some rubbers and plastics, to about 100 or so

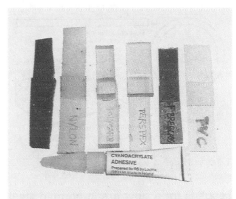

Some test pieces glued with cyanoacrylate 'instant' adhesive.

Gunmetal eccentric strap secured to faceplate using adhesive. Temporary joint while component is machined. (Photo courtesy Ted Jolliffe)

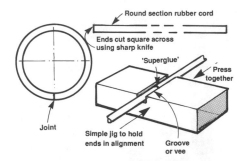

'O' ring splicing with instant adhesive.

seconds for a slow grade on mild steel or polycarbonate. Likewise the adhesive strength can vary from some 5 Newtons per square millimetre on neoprene or nitrile rubber, to between 20 and 30 Newtons per square millimetre tensile shear strength on mild steel.

As with other adhesives peel strengths are considerably lower than direct tensile or tensile shear values so joint design is important. A peel strength of about 4 Newtons per millimetre joint width is the best that can be achieved to date with a 'toughened

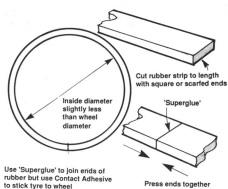

Making solid rubber tyres.

instant', but the normal grades will be quite a bit lower than this.

There is a tendency to regard any instant adhesive as the answer to all bonding problems, which is quite wrong. They have their limitations just like any other product and require the same care in their selection and use. Joint surface preparation is just as important and takes just as long as for other adhesives.

Although they can be used to bond a variety of materials to each other the strength of the completed joint will only be as great as the weaker of the two adhesive strengths. The main areas of use for instant adhesives is the rapid fixing of small metal or plastic parts and the bonding of small areas of rubbers, including the joining of 'O' ring materials.

STRUCTURAL ADHESIVES

We have seen in Chapter 8 that certain adhesives for wood not only give excellent adhesion they also maintain their adhesive strength under adverse environmental conditions. These properties render them suitable for building large highly stressed indoor and outdoor structures. The phenolic adhesives for instance were at one time employed to bond wood veneers for the construction of sailing boat hulls by the 'hot moulded' process. Such hulls were produced in heated moulds under pressure giving one piece frameless hulls of great strength and durability. They were, in effect, one large, fully shaped piece of marine grade plywood with no joints and no fastenings other than the adhesive. The advent of 'fibreglass', with the ability to cure at room temperature in cheaper single sided moulds without pressure, really made hot moulded hulls uncompetitive although

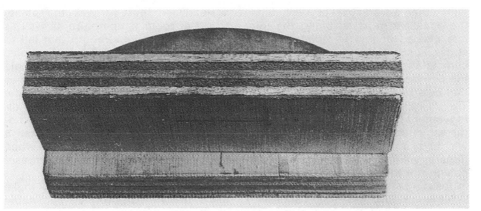

Exterior plywood glued between the laminations with waterproof adhesive.

the latter were in some ways superior to fibreglass.

Cold moulded wooden hulls have also been built of layers of wood veneers glued together with resorcinol/formaldehyde. The skins are laid over a wooden mould with adhesive brushed on between each layer. The pieces are kept in close contact by a multitude of metal staples during room temperature curing after which the staples can be removed. These are again one-piece hulls without joints equivalent to marine grade plywood.

Of course conventional flat sheets of marine ply can be used to construct hulls but they lack the rounded form that can be achieved by moulding and contain a mass of framing from which much of the strength is derived. There are many joint surfaces between frame components as well as between frame and skin which usually have permanent metal fastenings like screws or bolts, both to hold the parts together while the glue sets and to resist peel forces along the edges of the sheets.

Other structures are made from laminated timber beams. These are built up with layers of boards glued together and can be of complex curved and tapered shapes if desired. Buildings and bridges for example can be constructed in this manner to make best use of the stiffness, strength and durability which far exceeds that of solid timber of the same dimensions.

This then is what is meant by the term 'structural' adhesive. Strengths equivalent to that given by conventional mechanical fasteners are possible but

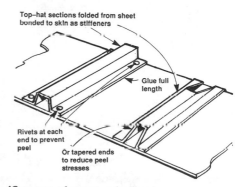

Top–hat sections folded from sheet bonded to skin as stiffeners

Glue full length

Rivets at each end to prevent peel

Or tapered ends to reduce peel stresses

One use of structural adhesives.

91

with other advantages as well. Since the adhesive grips all along the joint line instead of just at isolated points stress concentrations are reduced or eliminated. The adhesive also prevents the entry of moisture into the joints so that deterioration is less likely to occur. The continuous glue line prevents movement between parts, thus making the structure much stiffer for a given weight of material, and allows reductions in weight to be achieved by disposing the material to best advantage in resisting stresses.

Although the phenolic adhesives have been used to bond metals for many years (and are still used for this purpose in the bonding of brake linings to brake shoes), it was really the introduction of epoxide resins that made possible metal bonding on a large

scale. Once more it was the ability to set at room temperature as well as the high adhesive strength that led to widespread structural applications. Epoxides are produced as both single component and two-part products, as well as in solid hot-setting form; the latter is not to be confused with hot-melt adhesives, which are thermoplastics, as hot-set indicates that a chemical reaction takes place when heated resulting in a thermoset polymer.

The amateur is most likely to make use of the two-part products obtainable in squeeze tubes or syringes as slow setting or rapid set variations. The single pack type must be heated to enable setting to proceed.

The ability to form strong bonds with metals, wood and other materials

Tender frame fastened with rivets and epoxide adhesive.

means that epoxides are one of the most versatile structural adhesives available and are very easy to use. They also have excellent durability in severe conditions. The viscous nature of the mixed adhesive makes it suitable for large or complex assemblies as it will remain on the joint surfaces while parts are positioned and adjusted. Room temperature cure avoids the need for presses or heating arrangements.

The high viscosity does however make it more difficult to introduce the adhesive into restricted joint areas, including concentric fabrications, although gentle warming will reduce the viscosity to the point where it will flow. In this respect it may prove an advantage to gently warm small structures that have been assembled with adhesive and also rivetted, screwed or bolted. This allows the adhesive to flow into small gaps or crevices and seal around the fasteners to make liquid-tight joints. Another benefit is that neat fillets are formed at the joint edges improving appearance and lessening the possibility of peel separation by distributing stresses more widely. If intending to carry out these procedures the slower curing variety should be chosen to give ample time for positioning and rivetting before the heat is applied. As with other types of adhesive the setting time is influenced by the temperature of the surroundings; it begins to cure immediately after mixing, but the rate of cure depends upon the particular hardener and the temperature.

TOUGHENED STRUCTURAL ADHESIVES
These represent the latest development in the field of structural adhesives.

The term 'toughened' means Increased resistance to shock, vibration and peel stresses. This is brought about partly by variation of the polymer chemical structure and partly by the incorporation into the uncured adhesive of a fine dispersion of rubbery particles. Upon curing strong adhesive bonds are formed to the adherends and to the resilient particles which modify the properties of the more rigid adhesive polymer. Being in effect less brittle, the bond line can accommodate minute movements without fracture, which enables the stress to be spread over a wider area where it is safely absorbed. In addition any tiny cracks that are initiated in the adhesive layer are prevented from spreading by the presence of the resilient particles which act as crack-stoppers. In an ordinary brittle material cracks propagate very rapidly once they have been started so that failure is often catastrophic.

The two groups of these extra strong structural adhesives currently made are based on either acrylic or epoxide systems. The acrylic consists of one type which cures by the anaerobic method, and another type supplied in two-part form to be mixed before using, or the resin is applied to one of the joint surfaces and the hardener to the other so that they mix on contact. The toughened acrylics give high bond strengths with metals and most plastics but are not so good for some rubbers.

Toughened epoxides are probably the best all-round structural adhesives available today with excellent tensile shear strength allied to improved peel strength and the ability to withstand climatic conditions. Like the ordinary epoxies they have excellent adhesion to a wide range of substrates.

CHAPTER 10

MATCHING ADHESIVE TO MATERIAL AND CONDITIONS

ADHESIVES FOR PAPER AND CARD

Since these materials have little inherent strength there is not much point in using expensive adhesives capable of forming strong waterproof bonds. Paper and card are porous with an affinity for water, so normally one would choose a water-based adhesive for this application. Whichever one is selected will depend upon the areas to be covered, convenience of using, setting time, resistance to dampness, etc.

There are specialist products that have been developed to meet the needs of certain users, for instance in the office, model aircraft builders, papercraft enthusiasts, general household tasks and children.

The mounting of photographs needs some care as they are apt to buckle or stain if made too wet. Some products will stick them permanently while others allow for them to be detached fairly readily at a later date or for repositioning.

Listing a selection of products we have:

Evostick Paper Adhesive A pen type spreader for small areas.
Evostick Household Adhesive A sol-
vent-free product for general use, pen type.
Humbrol Tissue Paste For aircraft modellers and paper crafts.
Humbrol Gloy Liquid An office and household water-based adhesive.
Humbrol Gloy Paste A thicker version that does not run or penetrate too far.
Polycell Wallpaper Adhesive For large areas, powder mixed with water.
UHU Kids' Glue To stick paper, card, photos and wood.
UHU Stic Office and household glue that dries in a few minutes.
UHU Gluepen Water-based adhesive in a handy container.

The above are simple and safe products that give good bond strengths for the purpose but can be easily softened with water. More permanent bonds, or at least those more resistant to dampness, can be made with the following:

Copydex A rubber latex adhesive that dries to a film of rubber that does not soften in water.
Cow Gum A rubber solution that sticks to the surface rather than penetrating the paper – can be peeled off when required.

Evostick Multipurpose Clear A solvent type for a variety of uses.

Humbrol All Pupose Clear A solvent type that dries very quickly.

Pritt Stick (Henkel) A solid water-based stick in a lipstick style case for office and home use.

UHU All Purpose Adhesive A solvent type with many uses.

UHU All Purpose Extra A polyurethane type with a variety of uses.

In addition, any of the PVA adhesives generally known as 'white glue' may be used.

ADHESIVES FOR FABRICS

Where a non-permanent bond will suffice, any of the paper glues will work with fabrics manufactured from the natural fibres like cotton, silk, wool, etc. For more permanent results one of the specially developed fabric adhesives or the more water resistant types must be chosen. In those cases in which a fabric-to-fabric join is made a good degree of flexibility in the set adhesive is usually desirable, but when sticking fabric to wood, metal or other rigid substrates this is not so important, though the adhesive chosen must have an affinity for the substrate as well as for the fabric.

Since fabrics are porous materials, water-based or solvent adhesives may be chosen even when only one surface is fabric as it will enable the solvent to evaporate. Reactive adhesives can also be chosen if their ability to withstand more severe exposure conditions is a factor. This may also be necessary to ensure adhesion to some synthetic fibres.

The more permanent fabric adhesives are:

Copydex Rubber latex gives a clear flexible join.

Evostik Fabric Adhesive In a handy pen-style container.

Evostik Household Adhesive A general purpose product.

Evostik Multipurpose An impact adhesive of rubber/resin solvent type.

Evostik Multipurpose Clear A product giving a clear joint.

HMG AP Adhesive An all purpose impact type.

HMG Heat & Waterproof A clear adhesive usable up to 100 degrees Celsius.

Humbrol All Purpose Clear Quick drying and sticks to most surfaces.

Humbrol Extrabond A general purpose PVA adhesive.

UHU All Purpose A PVA solvent type adhesive.

UHU All Purpose Extra A polyurethane solvent type

Unibond Universal A PVA emulsion adhesive.

ADHESIVES FOR LEATHER

Leather is a natural substance whose degree of porosity depends upon the type of leather and the way it has been manufactured. It is of much less open structure than paper or fabrics but will, in most cases, absorb water-based or solvent adhesives.

There are really three different ways in which leather may need to be treated: (a) as a covering on small objects where little strength is required other than to hold it in place; (b) strong joints where movement is unlikely so a rigid adhesive can be used and (c) strong joints where a certain amount of flexibility in the joint is desirable. Examples of (b) would include items like fixing the heel of a shoe, while (c) might involve

Leather test pieces glued with various adhesives then broken by hand. Araldite-glued joint (left) could not be broken by hand.

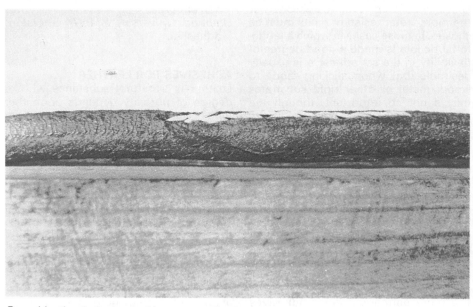

Round leather belt glued with contact adhesive and stitched.

repair of a watch-strap or the joining of round or flat leather belts.

Adhesives for group (a)

Evostik Leather Adhesive A special purpose formulation.

Evostik Multipurpose Impact Rubber/resin solvent type.

HMG All Purpose Adhesive An impact type solvent adhesive.

Humbrol All Purpose Clear Quick drying solvent type.

Any PVA adhesive Polyvinyl acetate emulsion or 'white glue'.

Adhesives for group (b)

Araldite Rapid A ten-minute-setting epoxide adhesive.

Araldite Standard A normal setting epoxide, usually overnight.

Araldite Superglue Wood and Leather A cyanoacrylate 'instant' adhesive.

HMG Epoxy Adhesive A normal setting epoxide.

Humbrol Superfast Epoxy A rapid setting epoxide.

Humbrol Wonderbond A cyanoacrylate 'instant' gel type.

Loctite Multibond 330 A two-part separate application adhesive.

Loctite Prism 401 A general purpose cyanoacrylate 'instant' adhesive.

Permabond E A two-part epoxide adhesive.

Permabond F A toughened acrylic adhesive.

UHU Supalok Gel A cyanoacrylate 'instant' gel type.

Adhesives for group (c)

This is the situation that requires most care in the selection of a suitable adhesive, since the material adjacent to the bond line can become impregnated with adhesive thus losing its flexibility.

Old joints should be scraped clean and wiped with degreasing solvent; new joints should be of scarf form made with a clean sharp knife. Belt drive joints should be stitched if an impact adhesive is used, otherwise the joint will 'creep' under sustained high loading. A reactive adhesive that is flexible when cured is the most satisfactory for this purpose.

For light loads, e.g. watch-straps, etc:

Evostik Leather Adhesive
Araldite Superglue Wood and Leather
   A cyanoacrylate 'instant' adhesive.
Humbrol Wonderbond A cyanoacrylate gel.
Loctite Prism 401 An 'instant' adhesive.
UHU Supalok Gel An 'instant' adhesive.

For sustained high loads:
Araldite Standard
HMG Epoxy Adhesive
Loctite Multibond 330
Permabond E

ADHESIVES FOR WOOD

There is probably a greater need to subdivide this section as these will be among the most used glues in the home workshop. Jobs can vary from replacing fragments of veneer on furniture, through the making of toys, boxes, cupboards, benches and other indoor items, to doors, windows, trailers or boats that have to be both strong and durable in all climatic conditions.

Where only wood-to-wood joints are concerned the adhesive chosen will depend upon the conditions under which it is to be used, but when the wood is to be fastened to another substrate the nature of that material must be taken into account too.

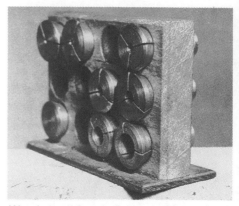

Wooden patterns glued with PVA adhesive. Fillets made with plastic wood.

Wooden collet rack. Contact adhesive fixed block to base.

No difficulty exists in finding adhesives that give bond strengths in excess of the strength of the wood itself. Some timbers, like balsa, may be joined by solvent adhesives whose quick drying is as important as the strength, though it should be noted that the ability to build up a fillet around the joint and so spread the load over a greater area is a very useful asset with such a soft material. Special purpose adhesives have long been available for such applications and it is wise to stick to them to ensure dependable results. Even with the hardest and strongest timbers most adhesives will, under dry conditions, give a bond line that remains largely intact under stress. Any failure is likely to involve breakage of the wood.

A first effort using balsawood.

Balsa-craft adhesives

HMG Balsa Cement
Humbrol Balsa Cement

Small repairs and small assemblies

Evostik Household Adhesive
Evostik Wood Adhesive PVA emulsion type.
Humbrol Carpenters' Wood Glue PVA emulsion type with aliphatic resins for extra tack.
UHU All Purpose Adhesive Solvent type.
Unibond Woodworker Adhesive PVA emulsion type.

Larger repairs and assemblies

Evostik Wood Adhesive
Humbrol Carpenter's Wood Glue
Unibond Woodworker Adhesive
HMG PVA Adhesive
Humbrol Cascamite Urea/formaldehyde type, one part, mix with water.

Small and large parts exposed to dampness

Evostik Wood Adhesive Waterproof For exterior and interior use.
HMG Watertite Adhesive A PVA co-polymer type.
Humbrol Cascamite Urea/formaldehyde powder, mix with water.
Humbrol Cascorez One-part cross-linking PVA adhesive.
Unibond Woodworker Waterproof A PVA co-polymer adhesive.

Small and large items exposed to very wet conditions

Araldite Standard Epoxide resin adhesive.
HMG Epoxy Adhesive
Humbrol Cascamite

Humbrol Cascophen Resorcinol/formaldehyde two-part adhesive.
Humbrol Superfast Epoxy
Permabond E Two-part epoxide adhesive.

Bonding decorative laminates to wood, plywood, chipboard

Evostik Impact Rubber/resin solvent type contact adhesive.
Evostik 528 Adhesive A contact adhesive.
Humbrol Cascamite For use in damp places. Work must be clamped until set.
Unilast A contact adhesive.
Humbrol Cascophen For very wet places. Clamp until set.

Bonding small metal or plastic items to wood

Araldite Rapid Epoxide resin adhesive.
Araldite Superglue Thick Gel Cyanoacrylate type.
Evostik Impact Contact adhesive.
HMG AP Adhesive Contact type.
HMG Epoxy Adhesive
Humbrol Superfast Epoxy
Humbrol Wonderbond Cyanoacrylate type.
Loctite Prism 454 Cyanoacrylate gel.
Permabond E Epoxy resin type.
Unilast Use as a contact adhesive.
UHU All Purpose Extra Polyurethane solvent adhesive.

Bonding larger metal or plastic parts to wood

Any of the epoxide adhesives
Loctite Multibond 330 A two-part separate application adhesive.
Permabond F A toughened acrylic adhesive.

Permabond ESP A single part toughened epoxy set by heat.

Special application

Humbrol Cascophen resorcinol/formaldehyde adhesive has been used to bond nylon fabric to wooden boat hulls as an abrasion resistant finish and to prevent penetration by marine wood-boring organisms.

Making wooden patterns from which metal castings may be produced is a job that many model engineers have to face at one time or another. Although simple shapes can often be carved out of a single block of wood, there is a danger that warping might alter the shape and dimensions or that thin sections might break away where end grain is encountered. For these reasons, as well as to simplify construction, it is usual to build the shape up from smaller pieces and glue them together; in this way the grain can be disposed to best advantage. Close grained, fine textured timbers are required for this work such as mahogany. Yellow pine was formerly much employed but seems difficult to obtain now. Good results have been obtained with PVA adhesives and the urea/formaldehyde type.

A special technique is used in the preparation of 'split patterns' to facilitate separation at the joint line after the shaping has been completed. The first step involves preparation of two pieces of wood corresponding to the thickness on either side of the joint. The surface of each piece is then coated with adhesive, but before the joint is closed a single sheet of newspaper is placed between the parts. After accurately aligning and clamping together the assembly is left until the glue is fully set. If the pattern is split on a cylindrical axis the glued block can be set up with one end in the lathe chuck with the jaws adjusted to bring the joint line central. For short pieces the work is then carefully turned to shape as far as possible,

Split wooden pattern for cylinder casting.

reversed in the chuck and the other end completed. For longer items tailstock support is necessary so the piece is first centre-drilled on the joint line before a small metal washer is placed on the tailstock centre prior to its being entered in the centrehole. The function of the washer is to prevent pressure on the point splitting the glue line.

When shaping is finished a small hole is drilled near each end at right angles to the joint and passing a short distance beyond it so that a pair of wooden or metal dowel pins can be fitted to locate the parts in the mould. Before fitting the pins a thin knife blade is tapped into the joint from one end causing the paper to split within its thickness so separating the two halves without damage.

Another prominent feature of casting patterns is the generous radius in the corners, around bosses and so on. In the old days shaped strips of leather 'fillet' were glued in these positions to form the curve, when painted it was indistinguishable from the wood. Now-adays it is usual to build the fillets up with some form of filler. Plastic wood is often used which can be purchased ready made or easily prepared by mixing fine sawdust or wood filings with a small quantity of adhesive to a thick paste immediately before use. An excellent product for the purpose is also Isopon car body filler which seems to shrink much less and is readily sanded smooth for painting. The cellulose fillers give a smooth finish in very thin layers so are useful for filling rough areas, but shrink excessively if applied in any thickness.

ADHESIVES FOR POROUS CERAMICS

For general use the requirement is

Unsaturated polyester resin/styrene based filler materials with good adhesive properties. Catalyst tube is inside lid of tins.

Glazed china object shattered; chipped glaze prevents neat repair.

Ceramic object with clean break.

usually for a colourless joint line with medium strength and moderate conditions of temperature and humidity.

Any PVA adhesive emulsion
Any cyanoacrylate gel 'instant' adhesive
Humbrol All Purpose Clear
UHU All Purpose Adhesive
UHU All Purpose Extra

Ceramic object repaired with Araldite epoxide adhesive giving an invisible joint line.

To withstand more severe conditions of temperature and humidity with increased strength.
Any epoxide adhesive
HMG Heat and Waterproof Adhesive

Adhesives for non-porous ceramics

For general use.
Any cyanoacrylate adhesive
For higher strength at increased temperature and humidity.
Any epoxide adhesive.

ADHESIVES FOR GLASS

Glass is difficult material to stick, particularly if it is to be exposed to very humid or wet conditions. It must be thoroughly degreased, abraded and dried before applying adhesive. Cyanoacrylates are only really suitable for light duties in dry conditions, the bond is likely to fail under humid conditions.

Silicone adhesives have good adhesion to glass but, because of their softness and flexibility, are more suited to sealing applications than to fastening.

Meter glass fastened to Bakelite case using silicone adhesive.

102

For high strength bonding of glass.

Any epoxide adhesive
Loctite Structural Adhesive 326
Permabond F Adhesive
Permabond VOX Adhesive

ADHESIVES FOR PLASTICS

Thermoplastic materials can often be dissolved by organic solvents therefore either the solvent alone may be used to make the bond or a solution of the same or a similar polymer.

Probably the most well known example to most people is the assembly of 'plastic kit' models which are available for a variety of subjects. Often the adhesive may be included with the kit, but specialist products have been designed for this market and include:
HMG Polystyrene Cement
Humbrol Polystyrene Cement Contains an 'anti glue-sniffing' agent.

Humbrol Liquid Poly For 'welded' joints, also with anti-sniffing agent.

These adhesives are very effective on the moulded polystyrene parts and of course can be used on other items whether made of plain or high impact polystyrene. A simple test to see if they are likely to stick the material is to place a dab on an inconspicuous part of the object, leave for a while then rub it off. If the surface has dissolved under the glue it should form a good joint, but remember that adequate time must be allowed after closing the joint for all the solvent to evaporate.

The Liquid Poly product is particularly good for the repair of hairline cracks in polystyrene articles as its low viscosity enables it to penetrate right through the break.

Where apparent failure to bond occurs it is wise to check that the material is in fact polystyrene – instances have been noted in which moulded kit parts of

A typical glued polystyrene model.

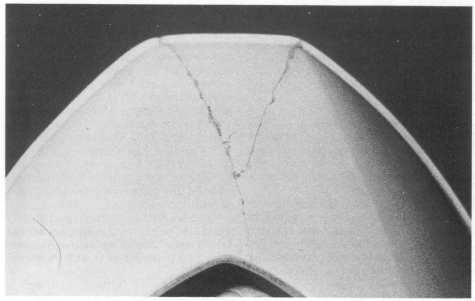

A polystyrene scale pan repair. The lower hair line crack was treated with Humbrol liquid poly and the loose piece was fixed with Humbrol polystyrene cement.

Perspex test pieces which were activated by briefly passing a gas flame over the surface then coated with epoxide resin adhesive. Fracture surface shows some pieces actually pulled out of the perspex.

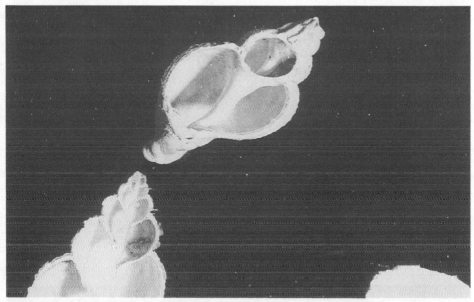

Sticking diverse materials. Sectioned sea shells stuck to perspex sheet with ICI Tensol No. 12 adhesive.

polythene have been used, for which polystyrene cements are quite unsuitable.

Perspex

A number of adhesive types can be used to bond this polymer. Of the common products the following have been found to give satisfactory bonds:

Humbrol Balsa Cement
Araldite Rapid Epoxide
Resorcinol/Formaldehyde

ICI produce several specialty adhesives, including *Tensol Cement No. 12* which is a solution of acrylic polymer with methyl methacrylate monomer in dichloromethane. Being a solvent adhesive it relies on absorption and evaporation of the solvent to cause setting, which therefore takes some time for full strength to be obtained.

Tensol Cement No. 70 is a two component type that sets by polymerisation at room temperature. This product gives high strength joints that are unaffected by outdoor exposure.

Cyanoacrylate adhesives give good results for small joint areas.

Polycarbonate

Humbrol Liquid Poly was found to give good joints with this material.

PVC

One of the most familiar forms in which this material is encountered must be rainwater gutters and piping, or bath and sink waste pipes. Some varieties are designed to be joined by fittings of

PVC pipe and elbow fitting with Gilflex-Key solvent-type adhesive.

A PVC gutter pipe section used as test piece for polystyrene adhesive, Isopon P38, and Araldite. They have been allowed to harden for 24 hours and have then been partly scraped away to check adhesion.

the same material fixed with a solvent-type adhesive. This may be obtained from DIY shops or builders' merchants as, for example, *Hunter Plastics PVC Cement*.

Another similar product may be obtained from electrical wholesalers as the adhesive used to joint PVC electrical conduit; an example of this solvent adhesive is *Gillflex-Key Solvent Weld Adhesive GW1*.

Besides their intended uses they may be employed for sticking all manner of PVC items.

Another product that has been found to adhere well to PVC is *Isopon P38* car body filler.

Nylon

Although rather difficult to stick in its natural state, nylon has been found to respond to the 'flame' surface treatment and can then be bonded by:
Evostik Contact Adhesive
Araldite Epoxide
Resorcinol/Formaldehyde

Acetal

The slippery surface of this material makes it very difficult for glues to adhere to, even after abrading, and it does not improve with the flame treatment. Of the adhesives tested only *Isopon P38* car body filler proved at all effective.

Polythene and polypropylene

These reputedly difficult materials are in fact relatively easy to bond to if thoroughly degreased then surface activated with a flame. Good results are obtained with:
Araldite Epoxide
Evostik Contact Adhesive
Loctite Prism 757 Polyolefin Primer

No visible means of support? Joint of 'flamed' polythene sheet glued with Araldite adhesive holding up heavy metal bars.

used with *Loctite Prism 406* is also effective without the flame treatment.

PTFE

Generally considered to be almost impossible to stick unless given very aggressive chemical surface treatment under industrial control, it has been claimed recently that a simple 'primer' treatment is effective:
Loctite Prism 757 Polyolefin Primer — the liquid is brushed over the joint surfaces and allowed to dry for 5 seconds

107

Workmate vice clamps with new rubber pads. Polypropylene jaws 'flamed' prior to coating with epoxide resin.

Loctite Polyolefin Primer and Prism 406 cyanoacrylate adhesive for joining inactive surfaces, e.g. PTFE and polythene or polypropylene.

or so before applying the cyanoacrylate adhesive *Loctite Prism 406*.

Thermoset polymers

These are generally easier to stick than the thermoplastics. Depending on the particular job, various adhesives may be used.

For large areas of sheet material in the form of decorative laminates etc use:
Evostik Contact Adhesive
Humbrol Cascamite Adhesive
Humbrol Cascophen Adhesive

For smaller areas and repairs:
Any epoxide adhesive
Any cyanoacrylate adhesive
Loctite Structural Adhesive 326
Loctite Multibond 330
Permabond F

ELASTOMERS

Natural rubber usually requires some sort of surface treatment to ensure a strong bond. This may be either thorough cleaning with trichlorethylene

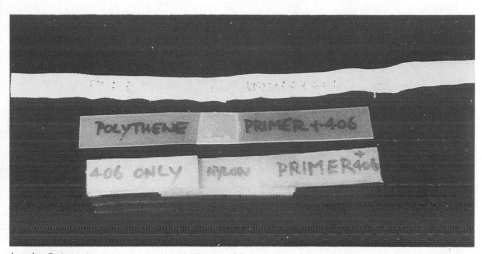

Loctite Polyolefin Primer used with Prism 406 cyanoacrylate adhesive on: top PTFE tape, centre Polythene, lower Nylon. The thin PTFE tape broke before the joint separated, but the adhesive failed with the thicker specimens.

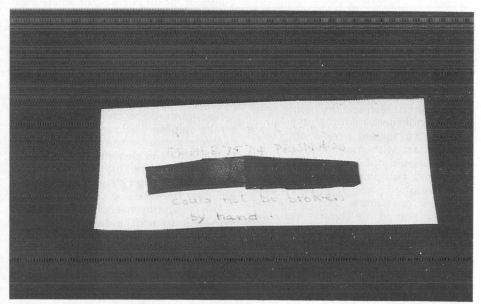

Natural rubber butt joint made with Loctite Polyolefin Primer and Prism 406 cyanoacrylate adhesive. Joint could not be broken by hand.

A joint in rubber that has been broken to show the epoxide adhesive still adhering to each surface.

solvent, complete surface abrasion of a 'scratching' nature, concentrated sulphuric acid treatment for several minutes following by washing, drying and flexing to finely craze the joint area, or treating the surface with household bleach to which a small quantity of very dilute hydrochloric acid has been added – apply to the joint area for a few minutes then wash and dry before coating with adhesive.

Note: the solvent should not come into contact with the hands and the vapour should not be breathed – best performed outdoors. If using concentrated sulphuric acid it must be kept in glass containers. Full protective clothing should be worn, including plastic gloves and goggles, and any spillage must be immediately washed off with plenty of water. The bleach solution gives off chlorine gas which is toxic. Wear protective gear, as described, and work outdoors. Flush spillage with plenty of water.

The adhesives which may be used on the rubber include solvent-type contact adhesive for large areas, or where high stresses are unlikely For highly stressed situations an epoxide adhesive is suitable, but where only small areas are involved an 'instant' adhesive is best.

Butyl rubber can be treated as for natural rubber, but the sulphuric acid method is not very effective.

Neoprene rubber can be treated best by the sulphuric acid method.

Nitrile rubber responds best to the trichlorethylene cleaning.

Fluoroelastomer may be treated as for PTFE, although it contains considerably less fluorine.

Ethylene/propylene co-polymer rubber responds to trichlorethylene cleaning and should respond to flame activation or to the Loctite Polyolefin Primer.

METALS

It is fortunate that adhesives form stronger bonds to mild steel than to other metals since it is the most common

Etched brass nameplate secured to wooden base with contact adhesive.

material in engineering use. The same adhesives can be used to join other metals, but the bond strengths will be different for each and all will be less than for mild steel.

As we have seen earlier, some form of surface treatment is desirable if consistently good results are to be obtained. Degreasing should always be performed followed by either abrasion or chemical etching. The metal should then be dried and the adhesive applied without delay.

Flat joints in sheet metal

Solvent type contact adhesive is suitable for those applications where little stress is involved. It is convenient for either small or large areas where the part has only to be held in position or support relatively light loads. With heavy loads the adhesive tends to 'creep', i.e. a slow stretching of the adhesive film takes place in which the parts may still be held together but relative movement occurs, particularly at higher temperatures. No precise figures for this behaviour have been found.

Cyanoacrylate instant adhesives provide a quick and easy, though expensive, way of fastening small components. They are generally usable at temperatures from about minus 50 to plus 80 degrees Celsius in the cured state and one or two up to 100 or even 120 degrees, but it should be remembered that the strength falls off at the upper end of the temperature range. Most instant adhesives also lose strength in conditions of high humidity. A drawback is their short shelf life of around six months, after which deterioration of performance or setting in the container is experienced, especially if it has been opened a number of times: this results in considerable wastage of an expensive product.

Products from the Loctite range have the following characteristics:

Prism 40 – general purpose fast cure, tensile shear strength on mild steel 18 to 26 N/mm^2 (1800 to 2600 N/cm^2 or 180 to 260 kgf/cm^2), on aluminium the strength would be 11 to 19 N/mm^2.

Prism 405 – toughened to give better peel strength and humidity resistance, tensile shear strength 22 to 30 N/mm^2 and 2 to 4 N/mm peel strength on mild steel.

Prism 499 – usable up to 120 degrees, tensile shear strength 18 to 26 N/mm^2 on mild steel, 11 to 19 N/mm^2 on aluminium.

Instant adhesives do require the joint to be well fitting as they have poor gap filling ability, usually not more than 0.1 mm.

Structural adhesives

Where maximum strength joints are

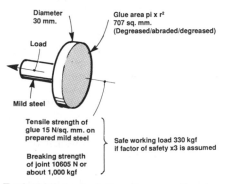

Diameter 30 mm.

Glue area pi x r²
707 sq. mm.
(Degreased/abraded/degreased)

Load

Mild steel

Tensile strength of glue 15 N/sq. mm. on prepared mild steel

Breaking strength of joint 10605 N or about 1,000 kgf

Safe working load 330 kgf if factor of safety x3 is assumed

Typical joint calculation direct tensile.

required in metal assemblies the choice usually falls on the epoxide adhesives, since these are widely available. The products on retail sale are two-part packs for mixing immediately prior to application, which will cure at room temperature or may be warmed to accelerate curing. In industry it is more usual to employ single-part adhesives in which the hardener and resin are combined during manufacture, but the rate of reaction at room temperature is extremely slow. It is necessary to heat this type, usually to a temperature of almost 200 degrees Celsius, to ensure rapid setting of the joint.

Common products which may be used by the amateur are typified by the two forms *Araldite Standard* and *Araldite Rapid*. The former may be mixed in either 100:80 or 100:100 resin to hardener proportions; the rapid variety is mixed in 100:100 ratio. The following data refers to these two products:

Pot life of Standard grade is about 100 minutes at room temperature; that of the Rapid grade is about 4 minutes at 20 degrees Celsius or 2½ minutes at 30 degrees.

A 0.05 to 0.10 mm thick glue line gives best performance with lap shear joints. Curing time for standard grade is 18 hours at 23 degrees, 3 hours at 40 degrees, 50 minutes at 70 degrees and 10 minutes at 100 degrees.

The Rapid grade sets in 1 hour at 23 degrees or ½ hour at 30 degrees.

Tensile shear strength of the Standard grade varies on different materials as follows: aluminium alloy – 18 N/mm^2, mild steel – 24 N/mm^2, stainless steel – 22 N/mm^2, copper – 23 N/mm^2, brass – 22 N/mm^2. The strength is constant at temperatures of minus 60 to plus 20 degrees, but then gradually falls to become about half the figures given when tested at plus 60 degrees.

Peel strength at room temperature is 5 N/mm joint width.

Shear strengths are little affected by petrol or lubricating oils, but kerosene (paraffin) and trichloroethane degrade the bond. Other solvents can cause considerable reduction of strength.

After 60 days in water at 20 degrees the strength is reduced to about half its original value while after 90 days at 90 degrees it is reduced to one sixth.

The fatigue strength of a lap joint loaded to 20 per cent of the static shear strength value is at least one million cycles at 90 Hz.

For the Rapid grade, shear strength with the above metals is more constant at around 14 N/mm² after 24 hours' cure at 23 degrees. The strength is improved by curing at 80 degrees for 2 hours.

Peel strengths are 3.5 N/mm width after 48 hours at 20 degrees and 5.5 N/N/mm width after curing for 2 hours at 80 degrees.

The reduction in strength caused by water immersion is less than for the Standard grade.

On the following materials the shear strength of a double lap joint 25 × 25 mm area, with 3 mm thick material, was:

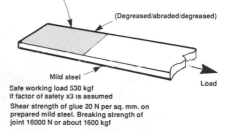

Joint dimensions 20 mm x 40 mm. Joint area 800 sq. mm.

(Degreased/abraded/degreased)

Mild steel

Load

Safe working load 530 kgf
If factor of safety x3 is assumed
Shear strength of glue 20 N per sq. mm. on prepared mild steel. Breaking strength of joint 16000 N or about 1600 kgf

Typical joint calculation tensile shear, single lap.

Anaerobic structural adhesives

These more recent introductions are based on acrylic resins and, like other anaerobics, the rate of cure varies with the materials being bonded. Some materials, like most metals, allow fairly rapid setting since they tend to speed up the rate of reaction; these are called 'active' surfaces. An exception is the purer grades of aluminium which are classed as 'inactive', although grades containing copper alloying additions are active. Non-metals are inactive too.

In order to speed up setting on inactive surfaces a liquid 'activator' solution is brushed or sprayed onto one of the inactive surfaces and the adhesive is applied to the other surface. When they are brought together setting com-

Acetal plastic	Tric. degreased	1.15 kN	breaking load with adhesive failure.
Polythene	Flamed	2.1 kN	breaking load material stretched.
Polythene	Sanded	880 N	breaking load adhesive failure.
Polypropylene	Flamed	1.83 kN	breaking load material stretched.
Polycarbonate	Degreased/abraded	3.8 N/mm²	adhesive failure.
Nylon	Degreased/abraded	3.8 N/mm²	adhesive failure.
Polyester/glass	Degreased/abraded	3.8 N/mm²	material failure.
ABS plastic	Degreased/abraded	3.4 N/mm²	adhesive failure.

All figures relate to Standard grade mixed 100:100 proportions.

Tightening side of sine table pivot bolt.

Pivot bolt of sine table fixed into boss of casting using high strength anaerobic retainer.

Slack magneto bearing race secured in housing with anaerobic retainer.

2¼ inch diameter bar secured in casting of milling machine using high strength anaerobic retainer.

Small boring head in which morse taper shank is held by high strength anaerobic retainer

Quorn tool and cutter grinder columns fastened to casting with high strength anaerobic retainer.

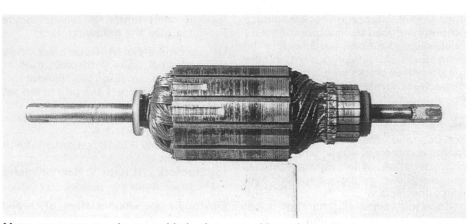

Motor armatures may be assembled using anaerobic retainer, adhesive tape on windings, and adhesive varnish for insulation.

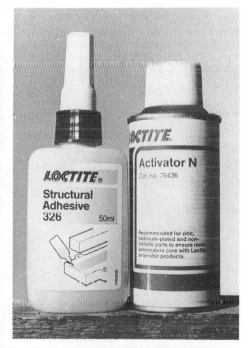

High strength structural adhesive and activator for metals, glass and plastics.

mences. Activators usually contain small amounts of copper compounds dissolved in a volatile solvent that takes only a few seconds to dry.

This group of adhesives is typified by *Loctite Structural Adhesive 326* for flat face bonding which gives a tensile strength of 15 to 25 N/mm² on grit-blasted mild steel. It will fill gaps up to 0.05mm. Glass, as well as other materials, can be bonded.

Loctite Multibond 330 is another version designed for use on metallic and non-metallic surfaces where larger gaps up to 0.40mm may exist. A tensile shear strength of 18 to 28 N/mm² on grit-blasted mild steel is claimed. Again an activator may be needed and usually 50 per cent of full strength is achieved within ten minutes of closing the joint.

Toughened structural adhesives

At present there are two types in this category, a toughened epoxide and a toughened acrylic. The main advantage of toughened adhesives is their improved resistance to shock loading, a property induced by incorporating very finely divided rubbery-particles into the resin. Such elastic components hinder the growth of cracks within the bond by dispersal of stress away from the crack tip.

Typical products in this category are the *Permabond 'F'* range of toughened acrylics and the *Permabond 'VOX'* range combining acrylic and epoxide characteristics. The *Permabond 'E'* range includes toughened epoxide two-part adhesives.

Good joint surface preparation is just as necessary when using structural adhesives as with any other type. In addition close fitting surfaces minimise gaps which are often a source of weakness. Large structures having a considerable length of joint require careful support and clamping during setting of the adhesive to obviate springing of the members. This applies particularly when sections are formed from thin sheet metal by bending into angle, channel or girder forms for bonding to sheet or other surfaces.

Structural adhesives are ideal for building up stressed skin or monocoque assemblies, or just for applying stiffening frames to flat panels. It is a wise precaution to rivet points at which peel separation is a possibility whether under normal stress conditions or overload, as in unforeseen impact situations.

Their use for fastening more massive components is exemplified by adoption of epoxide adhesive bonding for components of full size rail points and crossing assemblies.

CONCENTRIC JOINTS

The first to be considered are the thread locking compounds for which Loctite Products give the following data:

Using commercial M 10 size black oxide finished steel bolts with plain nuts in the unloaded condition the spanner torque required to break the grip of the set compound is:

for Screwlock 222 grade, 3 to 11 Newton metre (2 to 8 lb.ft.)

for Nutlock 242 8 to 18 Newton metre (6 to 13 lb.ft.)

for Studlock 270 16 to 36 Newton metre (12 to 27 lb.ft.)

Obviously for smaller sizes of thread the force required will be less and with the threads tensioned, the figure given would be added to the normal unscrewing torque. The figure of 3 Newton metre means that a force of 3 Newtons would need to be applied to the end of a spanner one metre long; force of 6 Newtons on a 50 cm spanner; 15 Newtons on a 20 cm spanner, and so on.

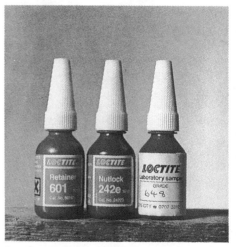

Loctite retainers: left high strength; centre low strength; right high temperature grade.

Studs secured in casting using threadlocking anaerobic compound.

The product used to retain bearing races in housings to restrain them from turning, as well as providing axial location, is Bearing Fit 641. Its static shear strength is 7 to 16 N/mm² (1015 to 2320 lbf/in²).

For assembling and permanently retaining cylindrical machine components there is Retainer 601 with static shear strength 16 to 30 N/mm². High Strength Retainer 638 static shear strength 22 to 40 N/mm².

These figures refer to assemblies in mild steel; for other metals the values given must be multiplied by these factors:

Cast iron 0.8 to 1.0

Copper and copper alloys 0.4 to 0.7

Aluminium 0.3 to 0.8 depending on its composition

Stainless steel 0.4 to 0.7

The Permabond firm has a similar range of products for these applications under their A 130 series for threadlocking, A 126, 148 and 168 for retaining.

Although anaerobic products are preferred for retaining duties because their low viscosity enables easy and convenient assembly, it is also perfectly satis-

factory to use epoxide adhesives where permanent fixing with high strength is desired. By gently warming a small quantity of the mixed adhesive, its viscosity is sufficiently reduced to aid placement even in relatively small assemblies. For this purpose the standard grade epoxide should be used since setting is more rapid at higher temperatures and the rapid grade would set too quickly.

ADHESIVE REPAIR PRODUCTS

Most people must now be aware of the car body repair products for application to badly corroded or accident-damaged metalwork. Besides such tasks they also find favour for filling and smoothing rough castings or welded joints prior to painting. Parts can also be built up, broken pieces replaced, etc, since the set material is capable of being sawn, filed, drilled and tapped. The same care

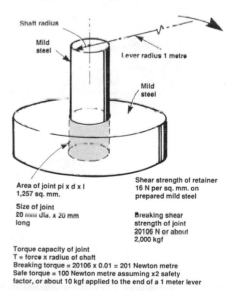

Shaft radius

Mild steel

Lever radius 1 metre

Mild steel

Area of joint pi x d x l
1,257 sq. mm.

Size of joint
20 mm dia. x 20 mm
long

Shear strength of retainer
16 N per sq. mm. on
prepared mild steel

Breaking shear
strength of joint
20106 N or about
2,000 kgf

Torque capacity of joint
T = force x radius of shaft
Breaking torque = 20106 x 0.01 = 201 Newton metre
Safe torque = 100 Newton metre assuming x2 safety
factor, or about 10 kgf applied to the end of a 1 meter lever

Typical concentric joint calculation.

Steel blocks glued with Isopon P38. That on the left required several sharp blows with a two pound hammer to break the joint.

Painted iron castings after 'filling' the surfaces with Isopon filler and rubbing down with fine abrasive paper.

must be taken in preparing surfaces as with other adhesive products, i.e. all rust and scale should be removed before the metal is abraded, traces of oil or grease must be washed off with degreasing solvent.

David's Isopon, being a typical product, consists of a two-part filled polyester resin with catalyst. The tin contains the resin while a tube of catalyst is fitted inside the plastic lid. The recommended mixing proportions for small jobs are to take a golfball size lump of resin with a bead of catalyst as big as a pea. These are mixed together to an even colour on a clean sheet of metal or plastic. It must be applied promptly as initial setting takes place within about five minutes, although some hours are required for it to reach full strength.

The polyester has good adhesion to a number of substrates including steel, aluminium, fibreglass, wood, etc, with a bond strength of about 1000 pounds per square inch for a mild steel lap joint (70 kg/cm²). Being a filled adhesive it has good gap filling properties which

is useful where joint surfaces may be somewhat rough. The same company's P 40 product is similar but filled with chopped glass fibres enabling it to bridge over gaps or holes.

Metal filled adhesives are also available for filling holes in castings, and building up worn or damaged parts with virtually solid metal. These are based on epoxide resins in which the two parts are mixed before use just as for the plain adhesives.

Araldite Filler for Metal has the filled resin and hardener in a double-barrelled plastic syringe for easy dispensing:
Loctite produce several grades:
Loctite Metal Set A1 filled with aluminium powder.
Loctite Metal Set S1 filled with steel powder.
Loctite Metal Set SS1 filled with stainless steel powder

Once fully hardened these products may be shaped by normal methods for solid metals, including drilling and tapping.

CHAPTER 11

SEALANTS

There are three basic ways which these products may be used: for sealing pipe threads and fittings; for use in place of gaskets to seal flanged joints; and to seal seams in structures thereby preventing leakage or entry of water.

(1) PIPE THREAD SEALANTS

Boss white has already been mentioned as being used for plumbing fittings on a domestic scale, although screwed joints have been dramatically reduced since the introduction of plastic pipe systems and with the increased use of capillary soldered copper joints.

Where the boss white is applied to such relatively large threads, it is customary to wind a few strands of hemp fibre into the thread grooves after they are coated to help fill the clearance spaces as the parts are screwed together.

It can be used on its own with smaller threaded fittings like those on models and is quite suitable for steam fittings. Where proper tapered threads are not used the joint should be screwed up tight or, if the fitting must be located at a certain angle, a locknut should be fitted to fasten it in the desired position.

More modern sealants are of the anaerobic type which set within the joint so will therefore both seal against leakage and lock the fitting in position. For the larger plumbing threads *Loctite Pipe Sealant 577* caters for sizes up to 50 mm OD. Being thixotropic it remains in place during assembly and cure. The force required to dismantle a cured joint of M10 thread size is 6 to 15 Nm (4 to 11 lb.ft.).

A stronger version for smaller threads up to 19 mm diameter is *Loctite Hydraulic Seal 542*. The dismantling force on M 10 size thread is 8 to 22 Nm (6 to 16 lb.ft.) These figures refer to tests in which the sealant was applied to a length of thread equal to the diameter; the dismantling force will of course vary with the size of thread.

(2) FLANGE SEALANTS

The modern tendency is to eliminate conventional cut gaskets, whether made from paper, cork, asbestos, etc, where possible and to rely on close fitting metal-to-metal surfaces with an anaerobic sealant to fill any minute gaps or scratches. *Loctite Multi-Gasket 574* is a typical product that will fill gaps up to 0.5 mm and operate at temperatures up to 150 degrees Celsius.

Thread-sealing adhesive used on cylinder connection of model hot-air engine. (Photo courtesy Ted Jolliffe)

It has a tensile strength on mild steel of 5 N/mm², a tensile shear strength on steel of 8 N/mm² and on aluminium of 5 N/mm².

Loctite 510 is a similar sealant usable up to 200 degrees Celsius, its tensile strength on steel is 8.5 N/mm², with tensile shear strength on steel of 9.5 N/mm² and on aluminium of 4 N/mm².

Loctite 573 has lower strength making it more suitable for joints that require to be dismantled at regular intervals. Tensile strength on steel is 5 N/mm², tensile shear strength on steel of 6.5 N/mm² and on aluminium of 3 N/mm².

These sealants are unaffected by engine oil at temperatures up to 125 degrees, but do lose strength in contact with petrol. At 22 degrees, which may be regarded as room temperature, the strength of 510 grade reduces to 80 per cent, of 574 to 75 per cent and of 573 to 60 per cent of the dry value previously stated.

There is a reduction in strength of 574 grade in contact with a water/glycol mixture at 87 degrees to 85 per cent and of 510 grade to 75 per cent. The 573 grade is unaffected.

Although a low pressure seal is obtained immediately after tightening the joint it typically takes up to 12 hours for full strength to be achieved.

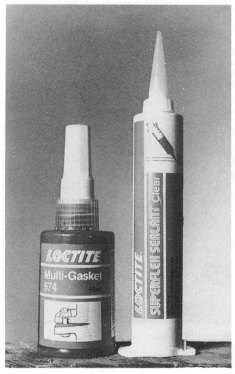

Liquid gasket products: left anaerobic, right silicone air-setting.

Other similar liquid gasket products are *Hermetite* and the *Permabond A 136*.

A different liquid gasket substance is the silicone type available as a thixotropic fluid which sets by reaction with atmospheric moisture. More familiar perhaps in its role as a DIY bath sealant, it has good adhesion to various surfaces but remains permanently flexible. The *Loctite Superflex Silicone Clear* has a tensile strength on mild steel of 1.9 N/mm², is usable from minus 60 to plus 250 degrees Celsius and resists moisture, steam and oil. This sealant is applied in a layer to one surface of the joint where it is allowed to set before closing the joint.

(3) STRUCTURAL ASSEMBLY SEALANTS

As already mentioned, adhesive bonded structural joints are automatically sealed against leakage and corrosion in addition to providing strength. Conventional bolted or rivetted joints may also be sealed by structural adhesives applied during assembly, in which case they improve the stiffness of the structure as well as performing the sealing function.

Where adequate strength is given by the mechanical fastenings being closely spaced or the thickness of the material providing sufficient stiffness to avoid deformation, it is possible to use the liquid gasket types of seal.

If an existing rivetted structure leaks it might be worthwhile trying one of the very fluid sealants such as *Loctite 290* penetrating grade which has the ability to creep into the joint before setting.

A point worth bearing in mind is that most grades of silicone sealant contain acetic acid; the 'vinegar' smell will be quite easily recognised. There is a danger that corrosion could arise from this source when using this type of sealant on ferrous or copper based metal parts. Either an engineering grade silicone stated to be suitable for metals, or a product stated to be free from acetic acid should be chosen if it is essential to avoid the possibility of attack. The *Evo-Stik Low Modulus Silicone*, for example, does not release acetic acid during cure, although some of their other grades do — it is important to read the data sheet before selecting any adhesive or sealant! Low modulus means that the set sealant is very flexible and elastic so that any movement in the joint will be taken up without rupturing the seal. This product is part of the range of Evo-Stik sealants designed for the building industry, but the data given shows that it could well be applied to model engineering problems in addition. For example, the life expectancy is 25 years plus, it will accommodate movement equal to the thickness of the joint, and it can be applied to metal or GRP panels as well as to other materials.

Many of these building sealants are designed for severe conditions of use so they should prove suitable for some of our requirements if carefully chosen. They are widely available in DIY stores, builders merchant's, etc.

A complete range of such items would include bitumen/rubber blends, acrylic elastomer, drying oil-based sealants, polysulphide rubber, etc, usually supplied in cartridges enabling blobs or strips to be squeezed out as required. Their main purpose is to act as barriers against moisture penetration, for instance the bitumen/rubber is intended for roof and gutter sealing, polysulphide for structural joints, panels, windows and so on. It would be necessary to carry out some simple tests under appropriate environmental conditions before finally deciding whether they might be applicable to a particular model function. Resistance to oil or fuel could easily be determined by allowing a thin film to set on a metal plate and then immersing it in the liquid for some period of time, possibly at an elevated temperature to speed up any effect. Many of us have these products to hand for household purposes so there is no reason why we should not attempt to adapt them for the hobby as well.

APPENDIX 1

Hints and Tips

(1) MOST IMPORTANT! Read manufacturer's recommendations and instructions *before* using the adhesive.

(2) Store adhesives at reasonably low temperature to ensure maximum shelf life – around 5 degrees Celsius is optimum for anaerobics, cyanoacrylates, etc, but do not let any adhesives freeze. Allow to warm to room temperature before use.

(3) Most adhesives are best applied to joint surfaces at normal room temperature of about 20 degrees Celsius. Avoid working at low temperatures as condensation may occur on joint surfaces and setting of adhesive will be much slower. Also evaporation of solvent adhesives is slow at low temperatures.

(4) Few adhesives retain significant strength when continuously used at temperatures exceeding 100 degrees Celsius. Take note of manufacturer's recommendations.

(5) If joint strength is required quickly either choose a rapid setting adhesive or warm the assembled joint (60 degrees Celsius is suitable for most adhesives), i.e. just too hot to be held in the hand.

(6) Hot humid conditions can seriously affect the bond strength of most adhesives, particularly on metal surfaces. If in doubt carry out tests before making joints in critical components.

(7) Consider adhesive bonding as an alternative to soldering, brazing or welding, where the heat of these operations might damage components, or where mechanical fastenings are unsightly or difficult to fit.

(8) Use adhesives to join different materials together, wood to metal, plastics to rubber, etc, as well as for joining similar materials.

(9) If unsure of adhesive suitability carry out a test on pieces of identical material prepared as for the job itself. Leave to fully set before checking bond strength.

(10) Where a single adhesive cannot be found to stick two dissimilar materials together, a 'three layer sandwich' may be made with a different adhesive used between each pair of surfaces.

(11) Do provide as much joint area as possible in relation to the loads

envisaged, and avoid sudden changes of material thickness at the joint line where it could lead to stress concentrations.

(12) Work out clamping arrangements and have everything prepared and ready to hand before starting to mix or apply glue.

(13) Take care to avoid trapping air when closing joints of any type since it reduces the area of contact leading to a weaker joint than expected.

(14) After abrading surfaces, clean off all dust or loose particles with a clean brush, by blowing with clean dry air, or by wiping with a clean cotton cloth dipped in appropriate solvent. It is often an advantage to scrub the adhesive into the surface using a clean stick to disperse loose particles and adequately wet the surface with glue.

(15) Use *clean* containers for mixing two-part adhesives. Well washed and dried plastic yoghurt, or other food, containers are suitable for most types.

(16) Stiff two-part adhesives like epoxides are best mixed on a clean flat surface such as freshly planed wood using a flat strip of wood or rigid plastic; thorough mixing is essential for success.

(17) With rapid setting epoxies or Isopon polyester it is most important to have everything ready prior to mixing because the adhesives begin to set in about five minutes. Joint strength will be very poor if applied too late.

(18) *Resist* the temptation to fiddle with a joint made with any adhesive just to see if it has set! Leave to stand for the recommended period of time so that setting will be complete.

(19) Adhesives may be used for making temporary joints as an aid to assembly, while drilling holes in matching parts and so on. It is usually sufficient to apply spots or strips of adhesive rather than cover the whole area which might make subsequent dismantling difficult. Adhesive tapes of various

Solenoid coil
secured with
adhesive PVC tape

Double-sided adhesive tape holding a heavy steel bar to a wooden support.

types are useful in this respect too.

(20) Where a machining mistake has occurred in concentric assemblies leading to a gap greater than about 2 thou, use epoxide adhesive rather than anaerobic because it sets throughout its bulk giving better gap filling.

(21) Filled adhesives tend to have better gap-filling abilities than most plain adhesives, so should be used where surface roughness is unavoidable but the joints are usually weaker than those made with well fitting surfaces.

(22) Rivetting a seam after an adhesive has set can cause cracking of rigid adhesive bonds through impact failure. It is best to rivet after applying adhesive but before it has set.

(23) While it is generally true that solvent-type adhesives are unsuitable for use on non-porous surfaces, the technique of using 'contact' or 'impact' adhesives should not be forgotten.

(24) Plastics are normally regarded as being non-porous yet some may be joined by solvent type adhesives. In this case the plastic must be capable of dissolving in the solvent, which becomes absorbed and eventually evaporates.

(25) It is true that good surface preparation is the key to successful adhesive bonding, but a few products have been developed that are 'oil tolerant'. *Loctite 603* for instance will retain oil-impregnated porous bronze bushes.

(26) Some solvent-type adhesives tend to form a skin over the surface if the joint is kept open too long. Work quickly to avoid this and twist the surfaces together if possible to break up any skin as the parts are put together.

(27) If really uncertain about an adhesives problem contact the relevant manufacturer for advice – they are usually very helpful. State your problem clearly giving full details of the materials, joint type proposed and conditions of operation.

APPENDIX 2

Useful Addresses

The following companies were contacted in connection with the preparation of this book and their help and advice is gratefully acknowledged.

Ciba-Geigy Plastics (*Araldite*)
Duxford
Cambridge
CB2 4QA

W David & Sons Ltd (*Isopon*)
Denington Industrial Estate
Denington Road
Wellingborough
Northants
NN8 2QP

Evode Ltd (*Evo-Stik*)
Common Road
Stafford
ST16 3EH

H Marcel Guest Ltd (*HMG*)
Riverside Works
Collyhurst Road
Manchester
M10 7RU

Henkel Adhesive Products (*Unibond*)
Road 5, Winsford Industrial Estate
Winsford
Cheshire
CW7 3QY

Humbrol Ltd (*Casco*)
Marfleet
Hull
HU9 5NE

ICI Chemicals and Polymers (*Tensol*)
P O Box 34
Darwen
Lancs
BB3 1QB

Loctite UK (*Loctite*)
Watchmead
Welwyn Garden City
Herts
AL7 1JB

Permabond (*Permabond*)
Woodside Road
Eastleigh
Hants
SO5 4EX

Rustins Ltd (*Shellac Varnish*)
Waterloo Road
Cricklewood
London
NW2 7TX

Sellotape Products (*Sellotape*)
Elstree Way
Borehamwood
Herts
WD6 1RU

UHU (*UHU*)
Brentford
Middlesex
TW8 9BD

A source of traditional animal glue for wood, as well as the glue pot to prepare it in, shellac sticks and a number of adhesives for wood, is:

Craft Supplies Ltd
The Mill
Millers Dale
Buxton
Derbyshire
SK17 8SN

Most of these products should be obtainable from DIY stores, builders' merchants, hardware stores, ironmongers, car accessory stores, engineering suppliers, or model shops and model engineers' stockists. In case of difficulty write to the manufacturer for advice.

GLOSSARY

Accelerator	A chemical that speeds up the action of a hardener.
Adherend	The material to be joined, the substrate.
Adhesion	Chemical bonding between adhesive and adherend.
Adhesive failure	Failure of joint within the glue line.
Adhesive strength	The force required to break a bond of unit area.
Anaerobic	In the absence of air
Assembly time	The time between applying adhesive and when it begins to set.
Bond	To join together with adhesive.
Bond strength	The same as adhesive strength.
Catalyst	(a) Strictly speaking is a substance that speeds up a chemical reaction without itself being used up.
	(b) For adhesives, is a chemical that initiates poly-merisation but is used up in the reaction.
Cohesion	The force holding atoms or molecules of a material together.
Cohesive strength	The bulk strength of a set adhesive.
Cold setting	An adhesive that will set at room temperature.
Contact adhesive	A solvent adhesive that is allowed to almost dry on both joint surfaces and adheres immediately they are in contact.
Cross-linking	A polymerisation reaction in which long chain molecules are joined to form a three-dimensional network.
Cure	To set by a chemical reaction.
Delamination	Failure of an adhesive joint allowing separation of layers.
Drying	Evaporation of a solvent from an adhesive.
Elastic	Capable of being stretched but returning to original dimensions when the load is released.
Elastomer	A polymer with pronounced elastic properties, a rubber.

Emulsion	A mixture of two immiscible liquids consisting of fine droplets of one suspended in the other.
Fatigue failure	Breakage caused by repeated application of stress, e.g. vibration.
Filler	An inert powder added to an adhesive.
Giant molecule	The complex three-dimensional structure formed by cross-linking reactions.
Hardener	One component of a two-part adhesive
Hot melt adhesive	Solid thermoplastic substance that is applied to the joint surfaces in a hot molten state and sets on cooling.
Inhibitor	A substance that prevents a chemical reaction from starting.
Instant adhesive	An adhesive that sets within seconds of the joint being closed.
Joint	The particular shape of the contacting faces where two parts are fastened together.
Joint area	Those areas of the joint coated with adhesive that contribute significantly to the strength.
Laminate	To glue together in layers.
Molecule	A number of atoms chemically joined together that forms the smallest unit of a substance.
Open time	The time during which an adhesive may be applied and the parts aligned before the adhesive begins to set. The same as assembly time.
Peel failure	The breaking of an adhesive joint by progressive opening from one end.
Peel strength	The force per unit width of the glue line required to separate the joint progressively from one end.
Plastic	Capable of being moulded into shape by pressure.
Plasticiser	A substance added to a polymer to improve its flexibility.
Plastics	Common name for synthetic polymers.
Polymer	A material formed by the chemical joining together of numerous small molecules of the same type to form much larger molecules.
Polymerisation	The chemical reaction whereby identical small molecules join to form much larger molecules.
Pot life	The time during which a reactive adhesive remains usable after mixing.
Primer	A chemical applied to a joint surface to improve the adhesion of an adhesive.

132

Priming coat	A preliminary application of a solvent adhesive to a very porous surface to seal the pores.
Quick setting	An adhesive formulation that sets more rapidly than the normal product
Resin	(a) One part of a two-part adhesive. (b) The sticky sap of pine trees.
Retainer	A type of adhesive designed to set in the gap between concentric components to prevent their movement under stress.
Rubber	A natural polymer having pronounced elastic properties.
Scarf joint	A tension joint with a diagonal glue line.
Separate application adhesive	A reactive two-part adhesive in which one component is applied to one joint surface and the other component to the mating surface. Mixing occurs on contact.
Setting	The hardening of an adhesive.
Setting time	The time during which an adhesive joint develops its full strength.
Shear strength	The force per unit area of contact required to break a lap joint by tension in the plane of the adhesive.
Silicone	An organic chemical in which the backbone of the molecular chain consists of silicon atoms joined together.
Solvent	A liquid that will dissolve other substances.
Stress	The force per unit area acting on a material.
Structural adhesive	Intended to be the major or only fastening of a structure or assembly and capable of carrying the loads imposed on the structure under severe environmental conditions.
Substrate	The material of the surface to be bonded.
Surface preparation	Preliminary operations carried out on joint surfaces prior to the application of the adhesive.
Surface treatment	Usually refers to a special chemical treatment of joint surfaces prior to application of adhesive to improve the bond.
Synthetic	Man-made substances.
Tack/Tackiness	The stickiness of a substance.
Tension	The stress in a material that is being stretched.
Thermoplastic	A material that is capable of being repeatedly heated and softened so that it might be reshaped.
Thermoset	A material that when heated becomes hard and cannot be softened and reshaped by further heating.
Toxic	Harmful to life.
Two-part adhesive	A chemically reactive adhesive that is supplied in two separate containers for mixing immediately before use.
Unvulcanised rubber	Raw rubber soluble in organic solvents.

Urethane polymers	Polymers formed by reacting di-isocyanates with polyhydric alcohols.
Viscosity	A measure of the 'thickness' of a liquid. Low viscosity liquids flow easily, high viscosity ones do not.
Volatile solvent	A solvent with a low boiling point that evaporates rapidly.
Vulcanised rubber	Rubber that has been heated with a chemical to cross-link the polymer chain molecules. No longer dissolves in organic solvents.
Wood failure	A broken joint where the adhesive bond is more or less intact but the wood itself has failed.

INDEX